Dynamics of Mechanical Systems

The direction in which education starts a man will determine his future life
Plato : *The Republic, Book V*
(427-347 BC)

It is a massy wheel,
Fixed on the summit of the highest mount,
To whose huge spokes ten thousand lesser things
Are mortis'd and adjoin'd; which when it falls;
Each small annexment, petty consequence,
Attends the boist'rous ruin.
Shakespeare, *Hamlet II, iii (1564-1616AD)*

"Talking of education, people have now a-days" (said he) "got a strange opinion that every thing should be taught by lectures. Now, I cannot see that lectures can do so much good as reading the books from which the lectures are taken. I know nothing that can be best taught by lectures, expect where experiments are to be shewn. You may teach chymestry by lectures. — You might teach making of shoes by lectures!"
James Boswell: *Life of Samuel Johnson, 1766* (1709-1784)

CARLISLE THOMAS FRANCIS ROSS, DSc

Carlisle Thomas Francis Ross was born in Kharagpur, India and educated in Bangalore at St Joseph's European High School during the closing years of the British Raj. Coming to England, he attended the Chatham Technical School for Boys (1948-51), followed by part time education at the Royal Dockyard Technical College (1951-56) at Chatham, Kent where he served a five-year shipwright apprenticeship. He proceeded to King's College, Newcastle-upon-Tyne (University of Durham), obtaining a B.Sc. (Hons) degree in naval architecture (1956-59); also working as a part-time draughtsman at HM Dockyard during university vacations.

For the next two years he laid the foundations of his powerful industrial experience as Designer in the Project Design Office at Vickers-Armstrongs (Shipbuilders), Barrow-in-Furness (1959-61). His outstanding work was acknowledged by promotion to the position of Deputy Chief of the Project Design Office. He next worked as a research graduate in the Department of Engineering, University of Manchester (1961-62), where in 1963 he gained his Ph.D. for research in stress analysis of pressure vessels.

He brought his industrial experience from Vickers-Armstrongs into teaching, first as Lecturer in Civil and Structural Engineering at Constantine College of Technology, Middlesborough (now University of Teesside) (1964-66), and later to Portsmouth Polytechnic (now University of Portsmouth) (1966-71) as Senior Lecturer in Mechanical Engineering, where he remains as Professor of Structural Mechanics.

His research research is based on computational methods, tested experimentally with colleagues. He has made important discoveries on the buckling of ring-stiffened cylinders and cones under external pressure, and has also developed the application of microcomputers on finite element analysis. In 1971 he co-invented the tube-stiffened pressure hull, making for greater submarine strength. In 1987 followed his structurally more efficient "bean can" pressure hull invention, and also the cheaper satisfactory dome-cup end for submarines. In 1995, with colleague David Jordan, came the invention of an important method for improving stability of roll-on roll-off car ferries, reducing risk of vessels capsizing in the event of flooding on the car deck. Currently he is developing a large drilling rig and a dredger, to enable oil and gas exploration in water to a depth of 5000 metres. This important application stems from his earlier research on pressure vessels when in Manchester University.

His contributions to engineering science were recognised in 1992 by the award of a Doctor of Science degree from the Council of National Academic Awards, London.

Dynamics of Mechanical Systems

Carl T.F. Ross, Bsc, PhD, Dsc, CEng, FRINA
Professor of Structural Dynamics
University of Portsmouth

Horwood Publishing
Chichester

First published in 1997 by
HORWOOD PUBLISHING LIMITED
International Publishers
Coll House, Westergate, Chichester, West Sussex, PO20 6QL England

British Library Cataloguing in Publication Data
A catalogue record of this book is available from the British Library

ISBN 1-898563-34-9

Printed in Great Britain by Hartnolls, Bodmin, Cornwall

DEDICATION

To my grandson, Edward Thomas Dryden, born 16th January 1997

Carl Ross and grandson, 15th February 1997

PREFACE

The book is aimed at first and second year undergraduate students in mechanical, civil and aeronautical engineering and naval architecture. It adopts a step-by-step methodological approach and addressses the mathematical difficulties faced by many modern day students. There are numerous worked examples and also a good supply of problems, with answers, on which the student can practice his/her newly acquired skills.

To the Student

Pay attention to the instructor and respect his/her knowledge. Do not talk while the instructor is talking; apart from the fact that is not polite, it will decrease the efficiency of the delivery of his/her lecture. Additionally, talking amongst each other may distract the attention of your peers. Work hard! Motivation is the biggest single factor in improving your ability. Remember that you can develop the potential of your mind by exercising it through study, rather similar to a body builder building up his muscles through physical exercise.

Do not be afraid to ask questions! There is an old Chinese proverb which states that "*it is better to ask a question and look a fool for five minutes than not ask a question at all and be a fool for the rest of your life.*" Remember also, that if you are working hard, then to keep stable, you must play hard; outside the lecture! One of the best ways of playing hard is to take regular strenuous physical exercise.

To the Instructor

There are many different ways that the topics covered in this book can be delivered to the students, and the instructor will have to find the best method that suits his/her style, within the constraints that he/she finds himself/herself. One such method which the author used successfully over a number of years was based on delivering the material contained in this book over sixty 1½ hour lecture/tutorial periods. Faced with this constraint and with a class size of about 30 students, the author found it satisfactory to divide the 1½ hour period into a ½ lecture, followed immediately by an hour tutorial.

The author's experience of teaching this and related topics over a period exceeding 32 years, was that most students can only pay full attention to a lecture on a numerate discipline for about 25 minutes, so what was the point of lecturing for one hour? Even if a student does not pay attention for 25 minutes, it is better to fulfil 25 minutes of his/her time than to waste a whole hour!

The author found no difficulty in coping with 30 or more undergraduates for the one hour tutorial period, providing all the students **started** with the same problem. The author would give the students hints on how to start the problem and leave them to get on with it for about 10 minutes. For students who '*did not have a clue*', he would advise them to study the theory that he had just covered. After this 10 minute period had elapsed, he would reveal on the board, part of the solution, so that the 'slower' student can catch up. He would then give the students hints on how to tackle the remainder of the problem and leave an appropriate time to achieve this. This process would be continued until the problem was eventually solved. In between the aforementioned short intervals, the author would give personal assistance to undergraduates who requested such assistance. Some of these students would make slow progress, but at least they would make some progress, while the 'faster' students would move at a faster rate. The author found an instructor's manual a valuable asset, as it enabled him to improve the efficiency of the delivery of his teaching.

The lecture was delivered with view foils and as the students had the book, they were able to follow the lecture and only take a few notes; the latter usually written in their course textbook. The tutorial was continued with a 'chalk and talk' approach, because 'chalk and talk' is more interactive for a numerate discipline. The author's view is that the purpose of the lecture was to teach the students 'what to learn' and the purpose of the tutorial was to teach the students 'how to learn it'.

This method worked very successfully for students with moderate and lower entry qualifications. The difference between heaven and hell is that '*in heaven you are faced with an infinite number of solvable problems and in hell you are faced with an infinite number of unsolvable problems.*'

Carl Ross
5th March 1997

ACKNOWLEDGEMENTS

The author would like to thank his wife Anne and his colleague Terry Johns for their assistance with the drawings and thanks are extended to Dr David Barton and David Richards for their helpful comments and advice.

Lastly, but not leastly, a thank you to Mrs Joanna Russell for her care and devotion in typing this book.

CONTENTS

			Page
Preface			
Acknowledgements			
Notation			
Chapter 1	Introduction		1
	1.1	Introduction	1
	1.2	Definitions	1
	1.3	Primary Units in Systeme International (SI)	2
	1.4	Some Simple Mathematical Formulae	2
	1.5	Dynamics	5
	1.6	Newton's Laws of Motion	5
	1.7	Newton's Laws of Gravitation	6
	1.8	The STROUD System of Units	7
Chapter 2	Kinematics of Particles		10
	2.1	Introduction	10
	2.2	Rectilinear Motion	10
	2.3	Plane Curvilinear Motion	19
	2.4	Rectangular Co-ordinates (x, y)	20
	2.5	Normal and Tangential Co-ordinates $(n - t)$	27
	2.6	Polar Co-ordinates $(r - \theta)$	37
	2.7	Relative Motion Using Translating Axes	45
	2.8	Constrained Motion of Connected Particles	55
Chapter 3	Kinetics of Particles: Force, Mass, Acceleration		62
	3.1	Introduction	62
	3.2	Application of Newton's 2nd Law of Motion	62
	3.3	Equations of Motion in Rectangular Co-ordinates	63
	3.4	Equations of Motion in other Co-ordinates	73
	3.5	D'Alembert's Principle	74
Chapter 4	Kinetics of Particles: Work, Energy and Power		84
	4.1	Introduction	84
	4.2	Work Done	84
	4.3	Elastic Strain Energy	86
	4.4	Kinetic Energy (KE)	88
	4.5	Power (P)	89
	4.6	Mechanical Energy (η)	89
	4.7	Potential Energy (PE)	89
	4.8	Conservation of Energy	90

		Page
Chapter 5	Kinetics of Particles: Momentum and Impulse	102
	5.1 Introduction	102
	5.2 Linear Impulse and Momentum	102
	5.3 Principle of Linear Momentum and Impulse	103
	5.4 Conservation of Linear Momentum	110
	5.5 Impact	114
	5.6 Angular Momentum	126
	5.7 Conservation of Angular Momentum	128
Chapter 6	Kinematics of Rigid Bodies	135
	6.1 Introduction	135
	6.2 Rigid Body Motion	135
	6.3 Translation	137
	6.4 Rotation about a Fixed Axis	138
	6.5 Absolute General Plane Motion Analysis	148
	6.6 Relative Motion Analysis: Velocity	153
	6.7 Instantaneous Centre of Zero Velocity	164
	6.8 Relative Motion Analysis: Acceleration	171
	6.9 Motion Relative to Rotating Axes	179
Chapter 7	Kinetics of Rigid Bodies	192
	7.1 Introduction	192
	7.2 Mass Moment of Inertia (I)	192
	7.3 Force: Mass: Acceleration	199
	7.4 Work, Power and Energy	231
	7.5 Impulse and Momentum	242
Chapter 8	Gyroscopic Theory	253
	8.1 Introduction	253
	8.2 Angular Momentum of a Rigid Body in Three Dimensions	254
	8.3 Gyroscopic Motion	256
Chapter 9	Free and Forced Vibrations	264
	9.1 Introduction	264
	9.2 Undamped Free Vibration	264
	9.3 Viscous Damped Free Vibration	277
	9.4 Damped Forced Vibrations	283
	9.5 Transmissibility	290
	9.6 Two and Multi-Degree of Freedom Systems	290
Appendix I	Simple Vector Analysis	294
Appendix II	Mass Moments of Inertia and other Geometrical Properties	297

NOTATION

$\bar{a}$	=	acceleration
$\bar{a}_{A/B}$	=	acceleration of 'A' relative to 'B'
$\bar{a}_G$	=	acceleration of the centre of gravity
a_n, a_t	=	accelerations in the 'n' and 't' directions, respectively
a	=	area
c	=	coefficient of viscous damping
c_c	=	critical damping factor
C	=	position of the instantaneous centre of zero velocity
$\bar{e}_n$	=	unit vector in the 'n' or normal direction
$\bar{e}_t$	=	unit vector in the 't' or tangential direction
$\bar{e}_r$	=	unit vector in the radial direction
$\bar{e}_\theta$	=	unit vector in the circumferential direction
e	=	the coefficient of restitution
E	=	Young's modulus of elasticity
f	=	the natural frequency (hz)
F	=	a force or a frictional force
g	=	acceleration due to gravity (normally taken as 9.81 m/s^2)
G	=	position of the centre of gravity or the universal gravitational constant
$\bar{G}$	=	linear momentum
h	=	height or distance
$\bar{H}$	=	angular momentum
$\bar{H}_G$	=	angular momentum about 'G'
$\bar{H}_o$	=	angular momentum about the point 'O'
$\dot{\bar{H}}_G$	=	the rate of change of the angular momentum with respect to the fixed frame of orientation
$(\dot{\bar{H}}_G)_{Gxyz}$	=	the rate of change of the angular momentum with respect to the rotating frame G_{xyz}
$\bar{i}$, $\bar{j}$, $\bar{k}$	=	unit vectors in the x, y and z directions, respectively
I, I_x I_y	=	mass moments of inertia (kg.m_2)
I_G	=	mass moment of inertia about the centre of gravity
I_o	=	mass moment of inertia about 'O'
I_{xy}	=	product of inertia
k	=	stiffness or radius of gyration
KE	=	Kinetic Energy
ℓ, L	=	length
m	=	mass
M	=	mass or moment, $\bar{m}$ = moment
M_G	=	moment about 'G'
M_o	=	moment about 'o'
MF	=	magnification factor
n	=	normal direction
N	=	normal component of reaction
P	=	force
PE	=	Potential Energy
$\bar{r}$	=	position vector
r^r	=	frequency ratio radius, $\bar{r}_A$, $\bar{r}_B$ = positional vectors of 'A' and 'B' respectively
$\bar{r}_{A/B}$	=	relative position of 'A' with respect to 'B'
R	=	frequency ratio radius, reaction
$\bar{s}$	=	position vector

s	=	displacement or distance
$\dot{s}$, $\ddot{s}$	=	velocity and acceleration, respectively
t	=	time or tangential direction
T	=	tension in the cord or rod
TR	=	transmittability ratio
$\overline{u}$	=	velocity
u	=	speed
$\overline{u}_A$, $\overline{u}_B$	=	velocity of 'A' and 'B', respectively
$u_{A/B}$	=	rotational velocity of 'A' relative to the fixed aims with 'B'
$\overline{u}_{rel}$	=	relative velocity of 'A' with respect to 'B'
$\overline{u}_G$	=	velocity of 'G'
WD	=	work done
x, y, z	=	rectangular co-ordinates
X, Y, Z	=	rectangular co-ordinates
$\dot{x}$, $\dot{y}$, $\dot{z}$	=	time derivatives of x, y and z
x_c, x_p	=	complementary function and particular integral, respectively
$\dot{\alpha}$, $\ddot{\alpha}$	=	angular velocity and angular acceleration, respectively
$\overline{\alpha}$	=	angular acceleration
δ	=	elongation or deflection
ζ	=	damping factor
λ	=	eigenvalue or characteristic root
η	=	mechanical efficiency
θ	=	angular co-ordinate
$\dot{\theta}$, $\ddot{\theta}$	=	angular velocity and acceleration, respectively
ϕ, ψ	=	phase difference
μ_k, μ_s	=	coefficients of kinetic and static friction, respectively
ρ	=	radius of curvature
ρ_m	=	density
τ	=	periodic time (seconds)
τ_d	=	periodic time for damped vibration (seconds)
$\overline{\omega}$, ω	=	angular velocity (rad/s)
ω_n	=	natural circular frequency
ω_d	=	damped circular frequency
$\overline{\Omega}$	=	angular velocity of the frame of reference

Parts of the Greek alphabet commonly used in mathematics

α	alpha
β	beta
γ	gamma
δ	delta
Δ	delta (capital)
ε	epsilon
ζ	zeta
η	eta
θ	theta
κ	kappa
λ	lambda
μ	mu
ν	nu
ξ	xi

Ξ	xi (capital)
π	pi
ρ	rho
σ	sigma
Σ	sigma (capital)
τ	tau
φ	phi
χ	chi
ψ	psi
ω	omega
Ω	omega (capital)

CHAPTER 1

Introduction

1.1 INTRODUCTION

This chapter commences with some basic definitions in dynamics. It is considered essential that the student familiarises him/herself thoroughly with these terms prior to moving onto the other chapters. This section is followed by a presentation of the primary units of dynamics, together with some basic formulae for use in dynamics and statics.

The chapter then presents Newton's three laws of motion and Newton's law of gravitation; this is followed by a presentation of the Stroud system of units. It is considered that a good grasp of units is essential for the student of engineering science and for this reason, examples are given on how to correctly manipulate units.

1.2 DEFINITIONS

A **scalar** is a quantity which has magnitude but no direction. Typical scalar quantities include length, time, mass and temperature.

A **vector** is a quantity which has both magnitude and direction. Typical vector quantities include force, weight, velocity and acceleration.

Mass (*m*) is a quantity of matter. It is a scalar and its units are in kilogrammes (kg).

Time (*t*) In Newtonian mechanics, time can be regarded as an absolute quantity of a sequence of events. It is a scalar and its units are in seconds(s).

Force (*F*) is a vector quantity and its units are in Newtons (N), where N = 1 kg.m/s^2.

Weight (*W*) is a special case of force. It is the gravitational attraction of a mass by the Earth. Its units are in Newtons (N).

A **particle** is a body whose dimensions can be regarded as negligible, so that its dimensions are not included in the dynamical analysis.

A **rigid body** is assumed to be a body whose dimensions do not change shape in a dynamical analysis.

The **displacement** of a particle can be regarded as the position of a particle at the time t_2 with reference to the time t_1. Displacement is a vector quantity and it can be positive or negative. Its units are in metres (m).

Velocity (*v*) is the rate of change of displacement with respect to time. It too is a vector.

$$\text{velocity} = v = \frac{ds}{dt}$$

$$= \frac{\text{displacement}}{(t_2 - t_1)} \text{ m/s}$$

where ds = displacement
$t_2 - t_1$ = an infinitesimally small time (very small).

Acceleration (*a*) is the rate of change of velocity with respect to time. It too is a vector.

$$\text{acceleration} = a = \frac{dv}{dt}$$

$$= \frac{v_2 - v_1}{(t_2 - t_1)} \text{ m/s}^2$$

where v_2 = velocity at time t_2
v_1 = velocity at time t_1
$t_2 - t_1$ = an infinitesimally small time (very small)

Momentum = mass x velocity (kg.m/s). It is a vector.

1.3 PRIMARY UNITS IN SYSTEME INTERNATIONAL (SI)

In general, in the SI system of units, it is best to use the primary units for problems involving dynamics, these are given in Table 1.1.

Table 1.1 Primary SI Units.

Quantity	Dimensional System	Unit	Symbol
Mass	*m*	kilogramme	kg
Length	*L*	metre	m
Time	*t*	second	s
Force	*F*	Newton	N

1.4 SOME SIMPLE MATHEMATICAL FORMULAE

The following mathematical formulae are of much use in statics and dynamics. Consider the right angled triangle of Figure 1.

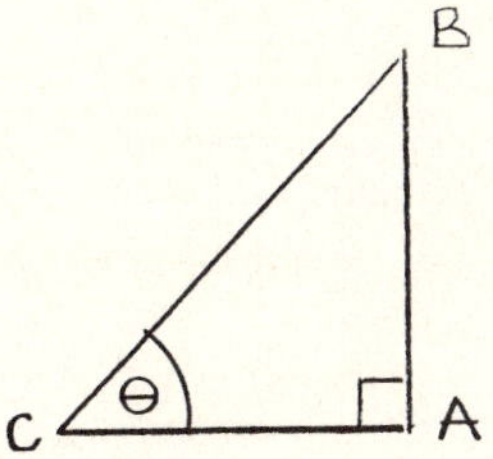

Figure 1.1 Right Angled Triangle.

From Figure 1.1

$$\text{Sin } \theta = AB/BC \quad (1.1)$$

$$\text{Cos } \theta = AC/BC \quad (1.2)$$

$$\text{Tan } \theta = AB/AC \quad (1.3)$$

The formulae of equations (1.1) to (1.3) can be used in simple vector analysis. A vector, such as a force, velocity or acceleration, say 'F', can be represented by two or more vectors, such as the vectors F_x and F_y of Figure 1.2.

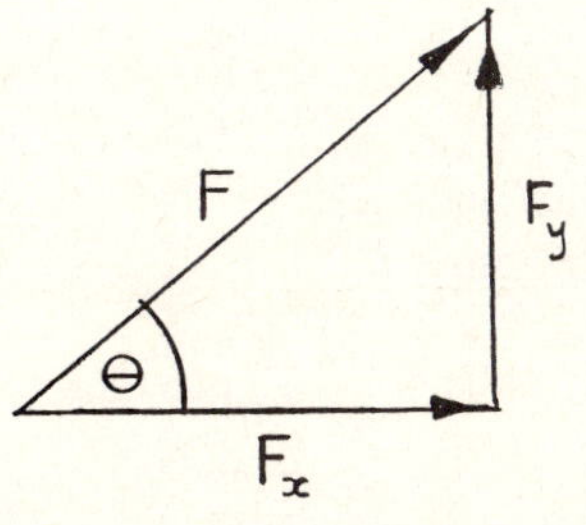

Figure 1.2 Vectors.

Now from equation (1.1)

$$\text{Sin } \theta = F_y/F$$

$$\text{or } F_y = F \sin\theta \quad (1.4)$$

That is, the vertical component of F, namely F_y, can be obtained from equation (1.4).

Similarly, from equation (1.2),

$$\text{Cos}\theta = F_x/F$$

$$\text{or } F_x = F \cos \theta \tag{1.5}$$

That is the horizontal component of F, namely F_x, can be calculated by equation (1.5).

NB Both F_x and F_y are also vectors.

In the case where the vectors cannot be represented by a right angled triangle then the sine and cosine formulae can be used;

These are defined with the aid of the triangle of Figure 1.3.

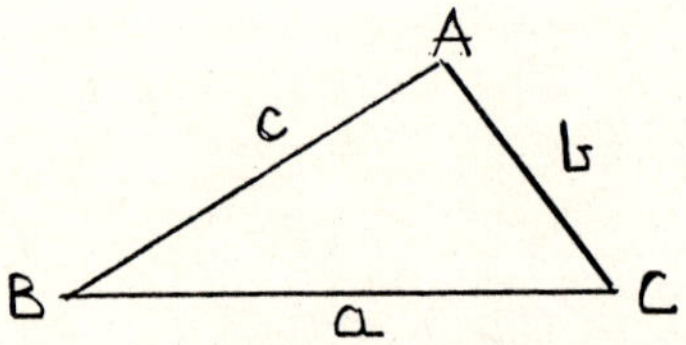

Figure 1.3 Triangle.

In the triangle of Figure 1.3,

A = ∠CAB
B = ∠ABC
C = ∠ACB

a = length of side opposite the angle A
b = length of side opposite the angle B
c = length of side opposite the angle C

With reference to Figure 1.3, the **sine rule** is given by equation (1.6).

$$\frac{a}{\text{Sin A}} = \frac{b}{\text{Sin B}} = \frac{c}{\text{Sin C}} \tag{1.6}$$

and the **cosine rule** is given by equation (1.7)

$$a^2 = b^2 + c^2 - 2\,bc \cos \text{A} \tag{1.7}$$

Depending on which quantities are known in Figure 1.3, either equation (1.6) or (1.7) can be used appropriately.

1.5 DYNAMICS

Dynamics can be regarded as that branch of engineering science which relates to the motion of particles and bodies. It follows on from statics, which is a study of the equilibrium of frameworks and bodies [1].

Dynamics consists of two sections, namely **kinematics** and **kinetics**; these are now defined.

Kinematics can be regarded as the study of the motion of particles and bodies, without reference to the forces causing the motion.

Kinetics is the study of the motion of particles and bodies with reference to the forces causing this motion.

Perhaps the father of dynamics was Galileo (1564 - 1642), who made a number of studies in dynamics, including that of free fall, the motion of a pendulum and the motion of a body on an inclined plane. A major breakthrough in dynamics was made by Sir Isaac Newton (1642 - 1727), who produced his three laws of motion and his law of gravitation; these are given below.

1.6 NEWTON'S LAWS OF MOTION

Newton's First Law states that 'Every body continues in its state of rest or uniform motion in a straight line, except if it is compelled by an unbalanced force to change that state'.

Newton's Second Law states that 'The rate of change of momentum, with respect to time, is proportional to the resultant force (F) and takes place in the direction of which this resultant force acts'.

Newton's Third Law states that 'To every action there is an equal and opposite reaction'.

The first law states that a state of rest or uniform motion can be treated in the same manner. An example of this is that of a fly flying in a compartment of a train travelling at a constant velocity. Despite the fact that the train may be travelling in a straight line at a uniform velocity of 160 km per hour, the fly does not get splattered against the compartment's bulkhead. In fact, it has no more difficulty in flying around the compartment of the train, regardless of whether the train is stationary or travelling at a uniform velocity.

If however, the train accelerates or decelerates, the fly's motion will be affected, as indeed will the comfort of the passengers. That is, if the train accelerates or decelerates, Newton's second law applies.

The third law applies when a force is applied to a body. In this case, if a force is applied to a body, the body in turn, resists this applied force with an equal and opposite reacting force.

There are of course a number of critics of Newton's laws of motion, but these laws are quite accurate providing the speed of the body is not too large. The reason for this is that if a body approaches the speed of light, its mass approaches infinity. The speed of light is 300,000 km/s or 186,000 miles/s. Additionally, according to Stephen Hawking[2], if the body reaches half the speed of light, the value of its mass increases by about 10%.

1.6.1 From Newton's Second Law of Motion

$$\text{Resultant force} = F = \frac{\mathrm{d}(mv)}{\mathrm{d}t} \tag{1.8}$$

where

m = mass

v = velocity

t = time

If the mass remains constant, equation (1.8) becomes

$$F = m\frac{\mathrm{d}v}{\mathrm{d}t} = ma \tag{1.9}$$

where

a = acceleration = $\mathrm{d}v/\mathrm{d}t$ (1.10)

1.7 NEWTON'S LAW OF GRAVITATION

Newton's law of gravitation states that the mutual attraction between two bodies is given by:

F = Attractive force between the two bodies

$$= \frac{G\, m_1\, m_2}{r^2} \tag{1.11}$$

where

m_1 = mass of one of the two bodies

m_2 = mass of the other body

r = distance between the centres of gravity of the masses m_1 and m_2

G = the universal gravitational constant

= $6.673 \times 10^{-11}\ \mathrm{m^3/(kg.s^2)}$ (1.12)

Now,

m_e = mass of the Earth

= 5.977×10^{24} kg (1.13)

and R = the mean radius of the Earth

$$= 6371 \text{ km} = 6.371 \text{ x } 10^3 \text{ m} \quad (1.14)$$

Hence for a mass of value 'm', which lies on the earth's surface, we have from equation (1.11)

$$F = \text{weight of } m = mg = \frac{Gm\,m_e}{R^2}$$

$$\text{or } g = Gm_e/R^2 = \text{acceleration due to gravity}$$

Substituting the values from equations (1.13) and (1.14) into equation (1.15), we can calculate a mean value for 'g' as follows:

$$g = \frac{5.977 \text{ x } 10^{24}\text{kg x } 6.673 \text{ x } 10^{-11}\text{m}^3/(\text{kg.s}^2)}{(6.371 \text{ x } 10^3)^2 \text{ m}^2} \quad (1.16)$$

$$= 9.826\text{m/s}^2$$

At **sea level**, $g = g_o = 9.807$ m/s² (1.17)

As the planet earth is an oblate spheroid, that is, it is flat in the polar regions, the value of 'g' will vary over the earth's surface.

$$\underline{g \text{ (equator)} = 9.780\text{m/s}^2} \quad (1.18)$$

$$\underline{g \text{ (Poles)} = 9.833\text{m/s}^2} \quad (1.19)$$

1.8 THE STROUD SYSTEM OF UNITS

In the calculation for 'g' from equation (1.16), it can be seen that where applicable, the units cancel, eventually leaving the correct units for 'g'. This cancelling of units is known as the Stroud system of units. Units are very important in engineering science, and it is the present author's experience that the bulk of student's errors are due to the improper use of units. The following examples will be used to demonstrate the Stroud system of units.

Example 1.1

Convert a pressure of 1 lbf/in^2 to Pa. It may be assumed that 1 in = 25.4 mm and 4.448N = 1 lbf.

$$\frac{1\text{lbf}}{\text{in}^2} = \frac{1\text{lbf}}{\text{in}^2} \text{ x } \frac{1\text{in}}{25.4\text{mm}} \text{ x } \frac{1\text{in}}{2.54\text{mm}} \text{ x } \frac{4.448\text{N}}{\text{lbf}} \tag{1.20}$$

$$= 6.894 \text{ x } 10^{-3}\text{N/mm}^2 = 6.894 \text{ x } 10^{-3}\text{MPa}$$

$$= 6.894 \text{ x } 10^{-3}\text{MPa x } \frac{10^6 \text{ Pa}}{1 \text{ MPa}} \tag{1.21}$$

or $\underline{1\text{lbf/in}^2 = 6894\text{Pa}}$

In equation (1.20), the 1 lbf/in^2 is first written down on the right hand side of the equation and then the appropriate conversion factors are used to convert lbf/in^2 to Pa, taking into consideration the cancelling process of the units. For example, to convert 'in^2' in the denominator to mm^2 in the denominator, it is necessary to multiply by (1 in/25.4 mm)^2, that is, the ins^2 cancel, leaving mm^2 in the denominator. It must be emphasised that by multiplying by (1 in/25.4 mm)^2 is the same as multiplying by one, because 1 in = 25.4 mm.

Similarly, to convert lbf on the numerator of the right hand side of equation (1.20), it is necessary to multiply the right hand side of the equation (1.20) by (4.448 N/lbf), so that the 'lbf' cancel, leaving 'N' in the numerator. Again, it must be emphasised that multiplying by (4.448 N/lbf) is the same as multiplying by one, because 4.448N = 1 lbf.

It can be seen that if this cancelling process of the units is used, fewer errors in the resulting units will occur.

Example 1.2

Convert density of lb/ft^3 to kg/m^3. It may be assumed that a mass of 1 lb = 0.454 kg and 1 ft = 0.3048 m.

$$\frac{1\text{lb}}{\text{ft}^3} = \frac{1\text{lb}}{\text{ft}^3} \text{ x } \frac{0.454\text{kg}}{\text{lb}} \text{ x } \frac{1\text{ft}^3}{(0.3048\text{m})^3}$$

or $$\frac{1\text{lb}}{\text{ft}^3} = \frac{16.03\text{kg}}{\text{m}^3}$$

Example 1.3

Convert a dynamic viscosity of 1 lbf.h/ft^2 to MN.s/m^2. It may be assumed that 1 ft = 0.3048 m and that a force of 4.448N = 1 lbf.

$$\frac{1\text{lbf.h}}{\text{ft}^2} = \frac{1\text{lbf.h}}{\text{ft}^2} \times \frac{1\text{ft}^2}{(0.3048\text{m})^2} \times \frac{3{,}600\text{s}}{\text{h}} \times \frac{4.448\text{N}}{\text{lbf}}$$

$$= 1.724 \times 10^5 \frac{\text{N.s}}{\text{m}^2} \times \frac{1\text{MN}}{10^6\text{N}}$$

or

$$\underline{1\text{lbf.hft}^2 = 0.1724\text{MN.s/m}^2}$$

Problems 1

1. A man weighs 800 N in London, where it can be assumed that g = 9.81 m/s^2. What does the man weigh in the equatorial region and also in the polar region, assuming that his mass remains constant?

   [Answer: 797.6 N, 801.9 N]

2. What does the man referred to in Problem 1 weigh on the moon, where g = 1.62 m/s^2, and what percentage would this weight be when compared with his weight in London?

   [Answer: 132.1 N, 16.5%]

3. Convert a distance of 10 km to miles. It may be assumed that 0.3048 m = 1 ft and 5280 ft = 1 mile.

   [Answer: 6.21 miles]

4. Convert 12 stone -5lbf to kgf. It may be assumed that 1 stone = 14 lbf and 2.2lbf = 1 kgf.

   [Answer: 78.64 kgf]

5. Convert a calorific value of 1 BTU/ft^3 to kJ/m^3. It may be assumed that 1 BTU = 1.055 kJ and 1 ft = 0.3048 m.

   [Answer: 37.26 kJ/m^3]

CHAPTER TWO

Kinematics of Particles

2.1 INTRODUCTION

This chapter is divided into two sections, one section is on rectilinear motion and the other section is on plane curvilinear motion. Now rectilinear motion can be described as one dimensional motion or motion in a straight line. The section on rectilinear motion is sub-divided into two sub-sections, namely motion when the acceleration is constant and motion when the acceleration varies with time or distance or velocity.

The section on plane curvilinear motion is divided into three sub-sections. One of these sub-sections is on plane motion with reference to the rectangular axes, namely x and y, while another sub-section is on plane motion with reference to the normal and tangential axes of the path of the motion. These axes are often called the n-t axes, where 'n' is perpendicular to the path of the motion and 't' is tangential to the path of the motion. Thus, if the path of the motion is curvilinear, the directions of 'n' and 't' will vary. The third sub-section on plane motion is on motion in terms of the polar co-ordinates 'r' and 'θ', where 'r' is radially outwards and 'θ' is tangential to 'r', in an angular direction.

The x-y rectangular co-ordinate system is popular when the motion or its direction is known in terms of the x-y co-ordinates. The n-t co-ordinate system is popular when the direction of the path that the motion follows is known. This may be in the case of a car or a train travelling around a curve, whose equation is known. Polar co-ordinates are popular for analysing mechanisms and other artefacts, which are rotating about a fixed point.

Also considered in this chapter is relative motion of translating axes and the motion of connected particles.

2.2 RECTILINEAR MOTION

Rectilinear motion of a particle can be described as motion in a straight line, as shown by Figure 2.1, where the particle 'P' moves to the position 'P^1' in time 'Δt'.

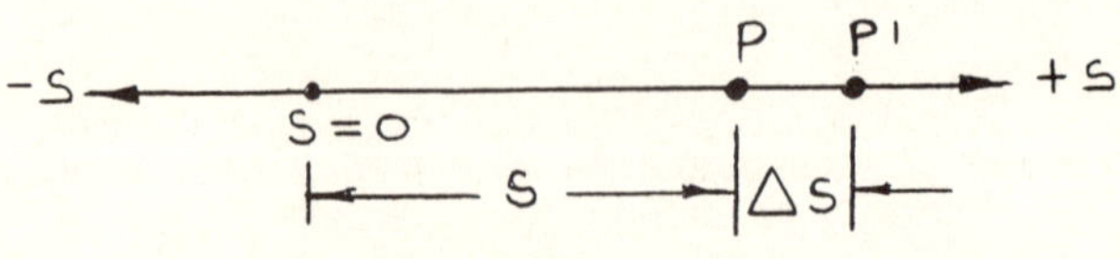

Figure 2.1 Rectilinear Motion.

In Figure 2.1, the distance 's' can be positive or negative. Consider the displacement Δs which takes place at the distance 's' and at the time 't', where Δs occurs over the time Δt.

Now from Chapter 1, the **velocity** υ is given by:

$$\upsilon = \frac{\Delta s}{\Delta t} \tag{2.1}$$

In the limit, when $\Delta t \to dt$ and $\Delta s \to ds$, the **instantaneous velocity** 'υ' is given by:

$$\upsilon = \frac{ds}{dt} \tag{2.2}$$

Similarly, we can obtain an expression for **acceleration**, when we consider Figure 2.2

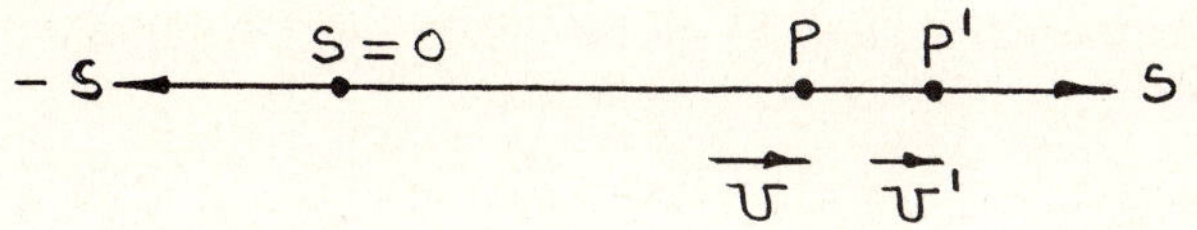

Figure 2.2

In Figure 2.2, the particle's velocity at the position 'P' is 'υ', when the time is 't' and at the position 'P^1' the particle's velocity is 'υ^1' at a time of $t + \Delta t$.

Hence the acceleration on 'a' is given by:

$$a = \frac{\upsilon^1 - \upsilon}{\Delta t} = \frac{\Delta \upsilon}{\Delta t} \tag{2.3}$$

where

$$\Delta \upsilon = \upsilon^1 - \upsilon$$

In the limit, when $\Delta t \to dt$ and $\Delta \upsilon \to d\upsilon$, the **instantaneous acceleration** 'a' is given by:

$$a = \frac{d\upsilon}{dt} \tag{2.4}$$

Now from equation (2.2)

$$\upsilon = \frac{ds}{dt}$$

$$\therefore a = \frac{d(ds/dt)}{dt} = \frac{d^2s}{dt^2} \tag{2.5}$$

The expression for the acceleration '*a*' can be put in another useful form by using the chain rule, as follows:

$$a = \frac{d\upsilon}{dt} = \frac{d\upsilon}{ds} \cdot \frac{ds}{dt}$$

$$\text{or } a = \upsilon \frac{d\upsilon}{ds} \tag{2.6}$$

These three expression for the acceleration '*a*', namely equations (2.4) to (2.6), all have their uses, depending on what is known about the motion.

2.2.1 Constant Acceleration

In this case the acceleration '*a*' is assumed to be a constant or zero. The assumption that '*a*' is a constant leads to a number of useful expressions for '*s*' and 'υ'; these are now derived.

From equation (2.4)

$$a = \frac{d\upsilon}{dt}$$

or $d\upsilon = a.dt$

$$\text{or } \int_{\upsilon_o}^{\upsilon} d\upsilon = \int_{o}^{t} a.dt = \text{area under } a\text{–}t \text{ curve} \tag{2.7}$$

As '*a*' is a constant, equation (2.7) becomes

$$\int_{\upsilon_o}^{\upsilon} d\upsilon = a \int_{o}^{t} dt$$

$$\text{or } [\upsilon]_{\upsilon_o}^{\upsilon} = a\,[t]_o^t$$

$$\text{or } \upsilon - \upsilon_o = at$$

$$\text{or } \upsilon = \upsilon_o + at, \tag{2.8}$$

where υ = velocity at the time 't', or final velocity
υ_o = velocity at when the time is zero or initial velocity
a = constant acceleration
t = the time taken to increase speed from υ_o to υ

Another useful expression can be obtained by substituting equation (2.2) into equation (2.8), as follows:

$$\upsilon = \frac{ds}{dt} = \upsilon_o + at$$

$$\text{or } \int_{s_o}^{s} ds = \int_{o}^{t} (\upsilon_o + at)\, dt = \text{area under } \upsilon - t \text{ curve}$$

$$\text{or } [s]_{s_o}^{s} = \left[\upsilon_o t + at^2/2\right]_o^t$$

$$\text{or } s - s_o = \upsilon_o t + \frac{1}{2} at^2$$

$$\text{or } s = s_o + \upsilon_o t + \frac{1}{2} at^2 \tag{2.9}$$

If the motion starts from $s_o = 0$, then equation (2.9) becomes

$$s = \upsilon_o t + \frac{1}{2} at^2 \tag{2.10}$$

where s_o = distance travelled when the time is zero
s = distance travelled in the time 't'

Another useful expression can be obtained from equation (2.6), where

$$a = \upsilon \frac{d\upsilon}{ds}$$

$$\text{or} \int_{s_o}^{s} a \, . \, ds = \int_{\upsilon_o}^{\upsilon} \upsilon \, . \, dv$$

$$\text{or} \left[a \, . \, s\right]_{s_o}^{s} = \left[\frac{\upsilon^2}{2}\right]_{\upsilon_o}^{\upsilon}$$

$$\text{or } a\left(s - s_o\right) = \frac{1}{2}\left(\upsilon^2 - \upsilon_o^2\right)$$

$$\text{or } \upsilon^2 = \upsilon_o^2 + 2a\left(s - s_o\right) \qquad (2.11)$$

If when $t = 0$, $s_o = 0$, then equation (2.11) becomes

$$\upsilon^2 = \upsilon_o^2 + 2\,as \qquad (2.12)$$

Equations (2.8) to (2.12) have a number of uses, depending on what is known about the motion of the particle. They are also useful for determining details of particles **falling under gravity**, when the effects of air resistance are neglected. In this case,

a = $-g$

where g = acceleration due to gravity usually taken as 9.81 m/s^2.

Example 2.1

A motor car starts from rest and travels along a straight and level road. The car accelerates uniformly for 10 s, reaching a velocity of 40 km/h. It then continues at this constant velocity for a further 50 s. How far does the vehicle travel over these two periods?

Consider the motion of the car in two stages, namely Stage 1 and Stage 2, and use the suffices 1 and 2 to represent these two stages.

Stage 1

$$\upsilon_o = s_o = 0$$

$$\upsilon = 40\,\frac{\text{km}}{\text{h}} \times \frac{1\text{h}}{3600\text{ s}} \times \frac{1000\text{ m}}{1\text{km}}$$

$$\underline{\upsilon} = \underline{11.11\text{ m/s}}$$

$$\underline{t_1} = \underline{10\text{ s}}$$

From equation (2.2), 'a' can be calculated

ie $a = (\upsilon - \upsilon_o)/t,$

$= (11.11 - 0) / 10$

$\underline{a = 1.111 \text{ m/s}^2}$

From equation (2.10)

$s_1 = 0 + ½ \times 1.111 \times 10^2$
$\underline{s_1 = 55.55 \text{ m}}$

Stage 2

$a = 0$

From equation (2.10)

$\underline{s_2 = 11.11 \times 50 + 0 = 555.5 \text{ m}}$

Total distance travelled $= s_1 + s_2$

$= 55.55 + 555.55$

$= \underline{611.10 \text{ m}}$

Example 2.2

A car is travelling at a velocity of 40 km/h, when it accelerates uniformly over a distance of 100 m to double its speed. Determine its acceleration over this period.

In this case, the following is known:

$$\upsilon_o = 40 \frac{\text{km}}{\text{h}} \times \frac{1\text{h}}{3600 \text{ s}} \times \frac{1000 \text{ m}}{\text{km}}$$

$= 11.11 \text{ m/s}$

$\upsilon = 80 \text{ km/h} = 22.22 \text{ m/s}$
$s_o = 0$
s = distanced travelled = 100 m

Equation (2.12) can be used as the only unknown in it is 'a'.

ie $\upsilon^2 = \upsilon_o^2 + 2as$
or $22.22^2 = 11.11^2 + 2 \times a \times 100$
or $493.73 = 123.43 + 200a$
or $a = 370.3/200 \text{ m/s}^2$
$\underline{a = 1.85 \text{ m/s}^2}$

Example 2.3

A train is travelling at 50 km/h, when it accelerates uniformly to a velocity of 100 km/h in a time of 45 seconds. What is its acceleration during this period?

In this case the following is known:

$$\upsilon_o = 50\ \frac{\text{km}}{\text{h}} \times \frac{1\text{h}}{3600\ \text{s}} \times \frac{1000\ \text{m}}{\text{km}}$$

$$= \underline{13.89\ \text{m/s}}$$

$$\upsilon = 100 \times 1000\ /\ 3600$$

$$= \underline{27.78\ \text{m/s}}$$

$$t = 45\ \text{s}$$

Equation (2.8) can be used as the only unknown in it is the acceleration 'a'

$$\text{ie } \upsilon = \upsilon_o + at$$

$$\text{or } 27.78 = 13.89 + a \times 45$$

$$\text{or } a = \frac{(27.78 - 13.89)}{45}$$

$$= \underline{0.309\ \text{m/s}^2}$$

2.2.2 Variable Acceleration

Examples 2.1 to 2.3 considered rectilinear kinematic problems involving only constant acceleration. However, a large number of problems in kinematics involve variable acceleration and some typical cases will now be considered.

Example 2.4

A small steel particle is released from rest, just above the surface of a fluid, where it is attracted by a magnet lying at the bottom of the fluid. If the steel particle descends through the fluid, according to the equation $s = (0.2\ t^3 + 2t)$ m, where,

s = the vertical position of the steel particle from the fluid surface (m),
t = time in seconds, from rest, determine the position, velocity and acceleration, when t = 2 seconds.

Now, $s = 0.2t^3 + 2t,$ (2.13)

so that when $t = 2$ s,

$s = 0.2 \times 2^3 + 2 \times 2$

$\underline{s = 5.6\ \text{m}}$

Now $\upsilon = \dfrac{ds}{dt}$

Hence, from equation (2.13)

$$\upsilon = 0.6\,t^2 + 2 \qquad (2.14)$$

Similarly,

$$a = \frac{dv}{dt}$$

Hence, from equation (2.14)

$$a = 1.2\,t \qquad (2.15)$$

Substituting **t = 2 seconds** into equation (2.14) and (2.15),

$$\upsilon = 0.6 \times 2^2 + 2$$
$$= \underline{4.4 \text{ m/s}}$$
$$\text{and } a = 1.2 \times 2$$
$$= 2.4 \text{ m/s}^2$$

Example 2.5

A rocket is launched vertically upwards from rest, and at a vertical distance 's' from where it is launched, its acceleration is defined by the equation $a = (4 + 0.01\,s)$ m/s^2. Determine its velocity when $s = 1$ km.

From equation (2.6)

$$a = \upsilon \,.\, \frac{d\upsilon}{ds} = 4 + 0.01\,s$$

$$\text{or} \int_0^{\upsilon} \upsilon \,.\, d\upsilon = \int_0^{1000} (4 + 0.01\,s)\, ds$$

$$\text{or} \left[\upsilon^2/2\right]_o^{\upsilon} = \left[4s + 0.01\,s^2/2\right]_o^{1000}$$

$$\text{or } \upsilon^2/2 = 4000 + 0.01 \times 1000^2/2$$
$$= 9000$$
$$\text{or } \upsilon = \sqrt{(9000 \times 2)}$$
$$= 134.2 \text{ m/s}$$

Problems 2.1

Neglect **air resistance** for all the problems in this section.

1. A ball is thrown vertically upwards from a distance of 2 m above the ground. If the ball is released with a velocity of 80 km/h, determine the height it will attain above the ground and the total time taken before the ball hits the ground.

   [27.12 m, 4.61 s]

2. A car starting from rest, accelerates with a constant acceleration of 2 m/s^2, until it reaches a required speed of 100 km/h. Determine the distance it travels from rest in a time of 2 minutes.

   [3.141 km]

3. A train is travelling at 120 km/h, when it is required to stop in a distance of 300 m. Assuming its deceleration is constant, determine the value of the deceleration that will be required to achieve this requirement.

   [1.85 m/s^2]

4. A ball bearing is dropped into a tank containing oil, so that its downward acceleration is governed by the equation $a = 2 - 0.1\ t^3$. Determine the time taken, distance travelled, and velocity when the acceleration is zero.

   [2.714 s, 7.012 m, 5.04 m/s]

5. A rocket is launched vertically upwards with an acceleration governed by the equation $a = 2 + 0.0001s$, where 's' is measured vertically upwards, in metres. If after travelling 200 m, its velocity is 200 m/s, what is its initial velocity?

   [16.3 m/s]

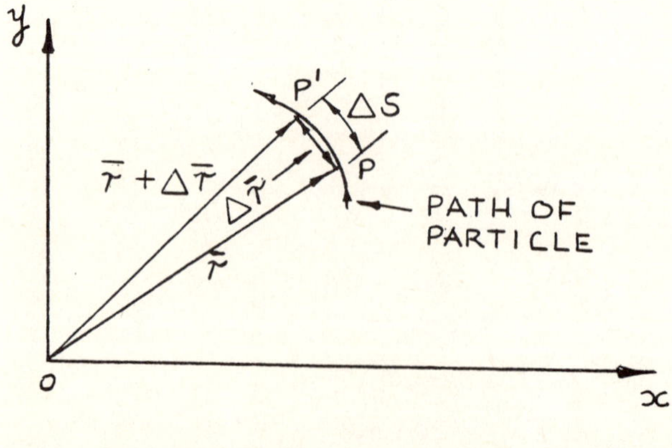

Figure 2.3 Plane Curved Motion.

2.3 PLANE CURVILINEAR MOTION

In this case, in order to derive the fundamental equations of motion, it will be necessary to consider motion in a curved path, about a fixed point 'O', as shown in Figure 2.3.

As the motion is in a plane, it will first be necessary to define position, displacement, velocity and acceleration, in terms of simple vector analysis, as follows:

2.3.1 Position

Consider the curvilinear motion of the particle 'P' about the fixed point 'O', where its position vector $\bar{r}$ is defined as:

$\bar{r} = x\,\bar{i} + y\,\bar{j}$, in rectangular co-ordinates,

where

$\bar{i}$ *and* $\bar{j}$ are unit vectors.

2.3.2 Displacement

Consider the displacement of the position vector $\bar{r}$ to the position vector $\bar{r} + \Delta\bar{r}$, as shown in Figure 2.3.

$$\text{Displacement vector} = (\bar{r} + \Delta\bar{r}) - \bar{r}$$

$$= \Delta\bar{r} \qquad (2.16)$$

2.3.3 Velocity

From Figure 2.3, the **average velocity** is

$$\bar{v}\ \text{average} = \Delta\bar{r} \,/\, \Delta t \qquad (2.17)$$

when $\Delta t \rightarrow dt$ and $\Delta\bar{r} \rightarrow d\bar{r}$, the **instantaneous velocity** $\bar{v}$ is given by:

$$\bar{v} = \frac{d\bar{r}}{dt} = \dot{\bar{r}} \qquad (2.18)$$

2.3.4 Acceleration

Consider again the curvilinear motion of the particle from P to P^1, as shown in Figure 2.4.

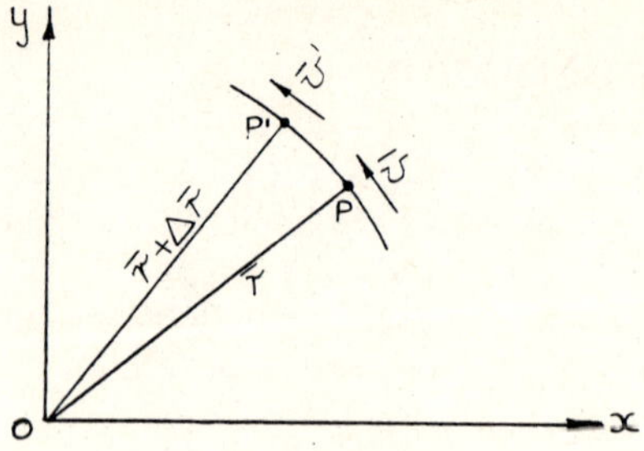

Figure 2.4

From Figure 2.4, it can be seen that the **average acceleration** is given by:

$$\bar{a}\ \text{average} = \overline{v}^{\text{I}} - \frac{\overline{v}}{\Delta t} = \frac{\Delta \overline{v}}{\Delta t} \tag{2.19}$$

When $\Delta t \rightarrow \mathrm{d}t$ and $\Delta\overline{v} \rightarrow \mathrm{d}\overline{v}$, the **instantaneous acceleration** is given by:

$$\bar{a} = \frac{\mathrm{d}\overline{v}}{\mathrm{d}t} = \dot{\overline{v}}$$

$$\text{but } \overline{v} = \frac{\mathrm{d}\overline{r}}{\mathrm{d}t} \tag{2.20}$$

$$\therefore\ \bar{a} = \frac{\mathrm{d}^2\overline{r}}{\mathrm{d}t^2} \tag{2.21}$$

2.4 RECTANGULAR CO-ORDINATES (x, y)

In terms of the rectangular co-ordinates x and y, as shown in Figure 2.3, the position vector $\overline{r}$ is given by:

$$\overline{r} = x\overline{i} + y\overline{j} \tag{2.21}$$

If $\overline{i}$ and $\overline{j}$ do not change with time, as is the case for rectangular co-ordinates, the **velocity** and **acceleration** vectors, namely equations (2.18) and (2.20), are given by:

$$\overline{v} = \frac{\mathrm{d}x}{\mathrm{d}t}\,.\,\overline{i} + \frac{\mathrm{d}y}{\mathrm{d}t}\,.\,\overline{j} \tag{2.22}$$

$$\text{and } \bar{a} = \frac{\mathrm{d}^2x}{\mathrm{d}t^2}\,.\,\overline{i} + \frac{\mathrm{d}^2y}{\mathrm{d}t^2}\,.\,\overline{j} \tag{2.23}$$

or

$$\bar{v} = \dot{x}\,\bar{i} + \dot{y}\,.\,\bar{j} \qquad (2.24)$$

$$\text{and } \bar{a} = \ddot{x}\,\bar{i} + \ddot{y}\,.\,\bar{j} \qquad (2.25)$$

or

$$\bar{v} = v_x\,\bar{i} + v_y\,.\,\bar{j} \qquad (2.26)$$

$$\text{and } \bar{a} = a_x\,.\,\bar{i} + a_y\,.\,\bar{j} \qquad (2.27)$$

where

$$v_x = \dot{x},\; v_y = \dot{y} \qquad (2.28)$$

$$\text{and } a_x = \ddot{x},\; a_y = \ddot{y} \qquad (2.29)$$

Now the resultant velocity of v is always a tangent to the path of the motion, as shown in Figure 2.5.

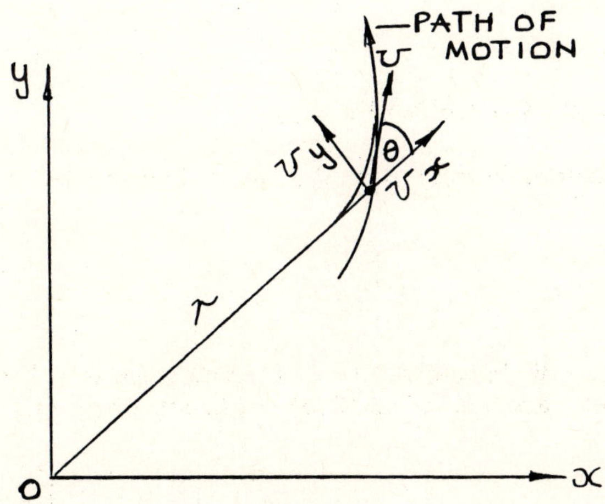

Figure 2.5

Hence, from Figure 2.5, it can be seen that the vector $\bar{v}$ can be resolved into the two vectors v_x and v_y, as shown above, so that from Pythagoras' theorem.

$$v^2 = v_x^2 + v_y^2$$

$$\text{or } v = \sqrt{\left(v_x^2 + v_y^2\right)} \tag{2.30}$$

$$\text{and } \theta = \tan^{-1}\left(v_y/v_x\right) \tag{2.31}$$

Also from Figure 2.5,

$$\tan\theta = \frac{dy}{dx}$$

$$\text{or } \frac{dy}{dx} = v_y/v_x \tag{2.32}$$

Hence from equation (2.32), if x and y are known as functions of 't', so that

$$x = f_1(t), \tag{2.33}$$

$$\text{and } y = f_2(t) \tag{2.34}$$

the equation for the path of the particle can be obtained in the form:

$$y = f(x) \tag{2.35}$$

Similarly the acceleration vector $\bar{a}$ can be resolved into two components, as shown in Figure 2.6.

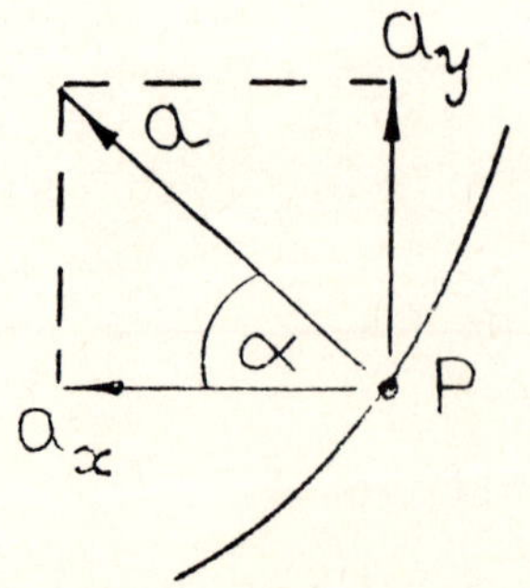

Figure 2.6

$$\text{ie } a = \sqrt{\left(a_x^2 + a_y^2\right)} \tag{2.36}$$

$$\text{and } \alpha = \tan^{-1}\left(a_y/a_x\right) \tag{2.37}$$

Example 2.6

A particle moves in a curvilinear motion according to the equation:

$$v_y = 40 - 9t \text{ m/s}$$
$$\text{and } x = 60 - 3t^2 \text{ m}$$

Plot the path the particle takes and determine its velocity and acceleration when x = 10 m. It may be assumed that when $t = 0$, $y = 0$.

$$\text{Now } v_y = \frac{dy}{dt} = 40 - 9t \tag{2.38}$$

$$\text{or } \int_o^y dy = \int_o^t (40 - 9t)dt$$

$$\text{or } y = 40t - 9t^2/2$$

$$\text{or } y = 40t - 4.5t^2$$

$$a_y = \frac{dv_y}{dt} = -9\text{m/s}^2$$

$$v_x = \frac{dx}{dt} - 6t \text{ m/s} \tag{2.39}$$

$$\text{and } a_x = \frac{dv_x}{dt} = -6 \text{ m/s}^2$$

Hence, from equation (2.36)

$$a = \sqrt{\left(a_x^2 + a_y^2\right)}$$

$$= \sqrt{(81 + 36)}$$

$$a = 10.82 \text{ m/s}^2$$

$$\text{when } x = 10\text{m}$$

$$10 = 60 - 3t^2$$

$$\text{or } 3t^2 = 50$$

$$\therefore t = 4.082\text{s}$$

Substituting this value of '*t*' into equations (2.38) and (2.39)

$$v_y = 3.262 \text{ m/s}$$

$$\text{and } v_x = -24.49 \text{ m/s}$$

Hence, from equation (2.30)

$$v = \sqrt{(24.49^2 + 3.262^2)}$$

$$v = 24.71 \text{ m/s}$$

Now

$$x = 60 - 3t^2 \tag{2.40}$$

$$\text{and } y = 40t - 4.5t^2 \tag{2.41}$$

Hence the path of motion of the particle can be obtained, by substituting suitable values of '*t*' into equations (2.40) and (2.41), as shown in Table 1.1.

Table 1.1

t	**0**	**1**	**2**	**3**	**3.262**
x	60	57	48	33	28.1
y	0	35.5	62	79.5	82.6

A plot of x against y is shown in Figure 2.7.

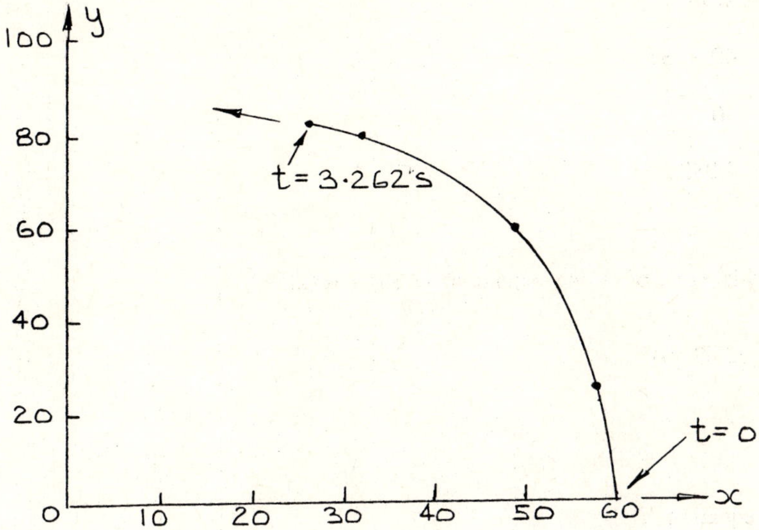

Figure 2.7 Path of Particle

Example 2.7

A projectile is launched horizontally from the top of a tower with a velocity v_o in Figure 2.8. If the projectile's range 'R' is 50 m and it takes 3 s for the projectile to reach the ground, determine the height of the tower and the value of the initial velocity 'v_o'. Neglect air resistance.

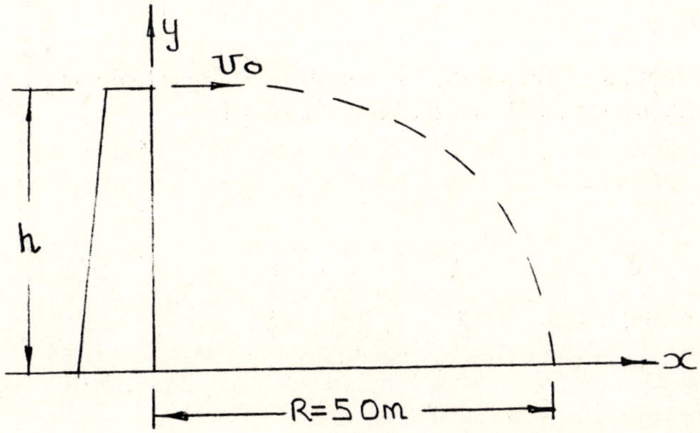

Figure 2.8

$$a_x = \frac{dx}{dt} = 0$$

$$\therefore v_x = \text{a constant} = v_o$$

$$a_y = \frac{dy}{dt} = -9.81\ \text{m/s}^2$$

$$v_y = 0 \text{ at } t = 0$$

From equation (2.10)

$$s = v_o t + \tfrac{1}{2} a t^2$$

Now in the vertical direction, $v_o = o$

$$\therefore y = -h = 0 - \tfrac{1}{2} \times 9.81 \times 3^2$$

$$\text{or } h = 44.15\ \text{m}$$

Also from equation (2.10)

$$x = v_o t + \tfrac{1}{2} a t^2$$

$$\text{or } 50 = v_o \times 3 + o$$

$$\therefore v_o = 16.67\ \text{m/s}$$

Problems 2.2

1. If the motion of a particle in rectangular co-ordinates is described by the equations

$$x = 2 + t$$
$$\text{and } y = 4t^2,$$

determine expressions for the velocity and acceleration in the x and y directions and obtain an expression for 'y' in terms of x.

[1 m/s, 8t m/s, 0 m/s^2, 8m/s $y = 4x^2 - 16x + 16$]

2. If the motion of a particle in rectangular co-ordinates is described by the equation $x = 27\,t^3$ and $y = 4t^2$, determine y in terms of x and the velocities and accelerations in the x and y directions, respectively.

[$y = 1.778\,x^{2/3}$, $81t^2$, $8t$, $162t$ m/s^2, 8m/s^2]

3. If the motion of a particle in rectangular co-ordinates can be described by the equations: $x = 2 \cos t$ and $y = \sin t$, determine expressions for y in terms of x and the velocities and accelerations in the x and y directions.

[$y = (16 - 4x^2)^{1/2}$, -2 sin t, 4 cos t, -2 cos t, -4 sin t]

4. A projectile is thrown at an angle θ to the horizontal. Determine the value of θ which will give the maximum range on level ground. Neglect air resistance.

[45°]

5. A projectile is fired on level ground at an angle θ to the horizontal. Determine the minimum possible muzzle velocity which will give a range of 10 km. What will be the maximum height attained?

[313 m/s, 3.536 km]

6. A ball is thrown at an angle of 30°, to the horizontal, down a slope of 30°, as shown in the Figure below. If the range along the slope is 80 m, determine the speed of delivery of the projectile v. What is the maximum height, 'h' above the point of delivery that the ball attains and the total time of flight of the ball.

[19.81 m/s, 5m, 8.155 s]

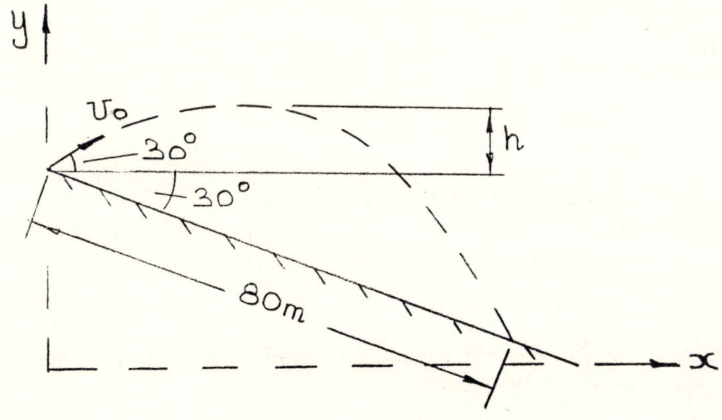

Figure 2.8 Motion Down a Slope

2.5 NORMAL AND TANGENTIAL CO-ORDINATES (*n-t*)

Another method of studying curvilinear motion is through the use of normal and tangential co-ordinates namely 'n' and 't' co-ordinates, respectively. In this case the symbol 'n' represents the normal co-ordinate and the symbol 't' represents the tangential co-ordinate, where these co-ordinates are with respect to the path of the motion. Thus, as the path of the motion is curvilinear, the directions 'n' and 't' will vary along the path, as shown in Figure 2.9.

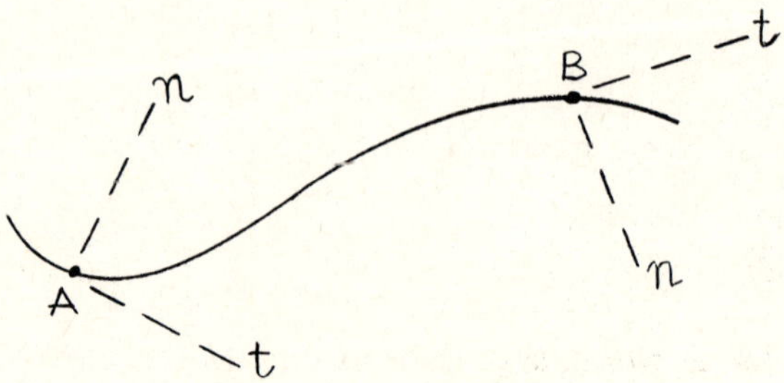

Figure 2.9 'n' and 't' Co-ordinates

It is evident from Figure 2.9 that the tangential velocity at any point, namely v_t, is given by:

$$v_t = v$$

where v = velocity of particle at any point

Similarly v_n = normal velocity at any point

$= 0$

Consider the motion of a particle, as shown in Figure 2.10, and let

$\overline{e_n}$ = a unit vector in the n-direction

$\overline{e_t}$ = a unit vector in the t-direction

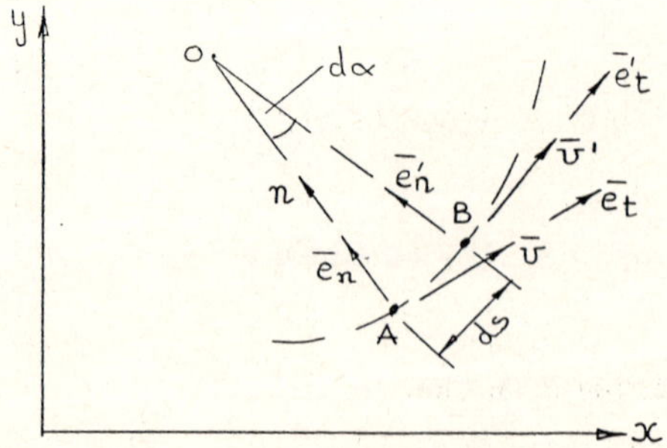

Figure 2.10

Now in Figure 2.10, the particle moves from A to B in the time dt, and during this small interval of time, let

ρ = radius of curvature of the path over AB, so that

$$\mathrm{d}s = \rho \, \mathrm{d}\alpha \tag{2.42}$$

where α is in radians

$$\text{Now } v = \frac{\mathrm{d}s}{\mathrm{d}t} = \rho \frac{\mathrm{d}\alpha}{\mathrm{d}t} \tag{2.43}$$

hence, the velocity vector can be written as

$$\bar{v} = v\,\bar{e}_t = \rho\,\dot{\alpha}\,\bar{e}_t \tag{2.44}$$

Now

$$\bar{a} = \frac{\mathrm{d}\bar{v}}{\mathrm{d}t} = \frac{\mathrm{d}\left(v\bar{e}_t\right)}{\mathrm{d}t} \tag{2.45}$$

$$= v\dot{\bar{e}}_t + \dot{v}\,\bar{e}_t$$

To obtain $\bar{e}_t$, consider Figure 2.11

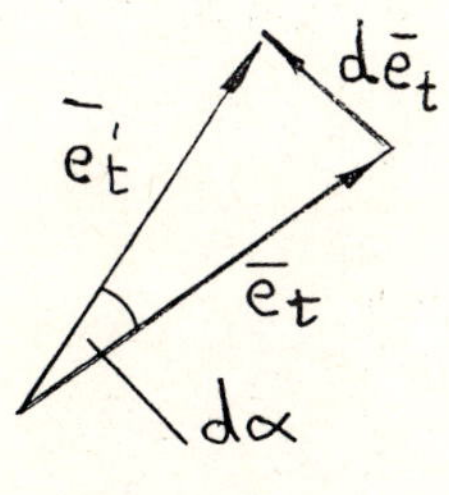

Figure 2.11

From Figure 2.11,

$$\bar{e}_t^{\,1} - \bar{e}_t = \mathrm{d}\bar{e}_t$$

but the vector d$\bar{e}_t$ is in the direction $\bar{e}_n$, so that

$$\mathrm{d}\bar{e}_t = \bar{e}_n \, \mathrm{d}\alpha \tag{2.46}$$

and $$\frac{d\overline{e}_t}{dt} = \overline{e}_n \frac{d\alpha}{dt} \tag{2.47}$$

Hence

$$\overline{a} = v \,.\, \overline{e}_n \,.\, \frac{d\alpha}{dt} + \dot{v}\overline{e}_t \tag{2.48}$$

but from equation (2.43)

$$\frac{d\alpha}{dt} = v/\rho, \tag{2.49}$$

so that

$$\overline{a} = \frac{v^2}{\rho} \,.\, \overline{e}_n + \dot{v}\,\overline{e}_t \tag{2.50}$$

ie

$$\begin{aligned} a_n &= \text{normal acceleration to the path} \\ &= v^2/\rho \end{aligned} \tag{2.51}$$

and

$$\begin{aligned} a_t &= \text{tangential acceleration to the path} \\ &= \dot{v} = \frac{dv}{dt} = \frac{d^2 s}{dt^2} \end{aligned} \tag{2.52}$$

Another useful expression for a_t is as follows:

$$\begin{aligned} a_t &= \frac{dv}{dt} = \frac{d(\rho\dot{\alpha})}{dt} \\ a_t &= \rho\ddot{\alpha} + \dot{\rho}\dot{\alpha} \end{aligned} \tag{2.53}$$

For **circular motion**, $\rho = r$ = the radius of the turning circle = a constant, so that

$$v = r\dot{\theta} \tag{2.54}$$

$$a_n = v^2/r = r\dot{\theta}^2 = v\dot{\theta} \tag{2.55}$$

$$a_t = \dot{v} = r\ddot{\theta} \tag{2.56}$$

where

$\theta = \alpha$

$\dot{\theta} = \dot{\alpha}$ See Figure 2.12

and $\ddot{\theta} = \ddot{\alpha}$

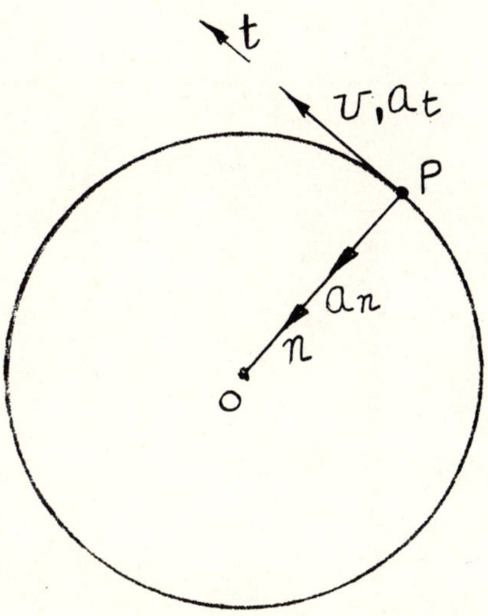

Figure 2.12 Motion in a Circle

where

$\dot{\theta}$ = angular velocity

$\ddot{\theta}$ = angular acceleration

a_n = centripetal acceleration

Example 2.8

The equation for a ski slope is given by $y = x^2/30$, where x and y are defined in Figure 2.13. If a skier is travelling along this path at a constant speed of 5 m/s, what is his velocity and acceleration when he arrives at the point A. Assume that the skier's dimensions are small, so that he could be regarded as a particle.

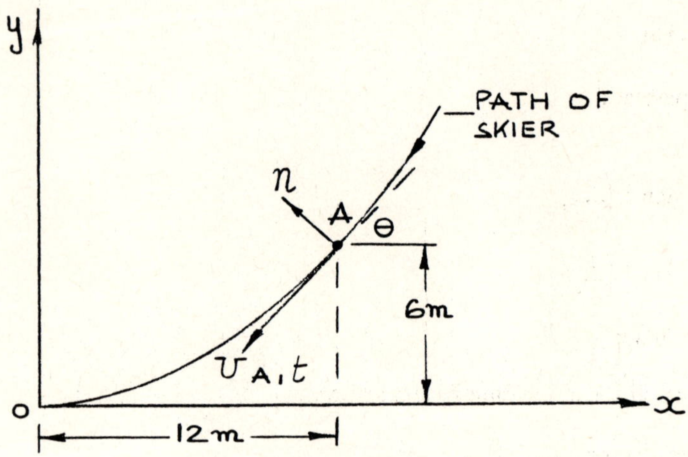

Figure 2.13 Path of Skier

$$y = x^2/30$$

$$\frac{dy}{dx} = \frac{2x}{30} = \frac{x}{15}$$

$$\frac{d^2y}{dx^2} = \frac{1}{15}$$

$$\text{Now, } \rho = \left\{\frac{[1 + (dy/dx)^2]^{\frac{3}{2}}}{d^2y \,/\, dx^2}\right\} \tag{2.57}$$

= radius of curvature

$$\text{ie, } \rho = \frac{[1 + (x/15)^2]^{\frac{3}{2}}}{1/15}$$

At $x = 12\text{ m}$,

$$\rho = [1 + (12/15)^2]^{\frac{3}{2}} x\ 15$$
$$= 1.64^{1.5} x\ 15$$
$$\rho = 31.5\text{ m}$$

Inclination of path at A = θ,

$$\text{where } \theta = \tan^{-1}\left(\frac{dy}{dx}\right)$$
$$= \tan^{-1}\left(\frac{12}{15}\right)$$
$$\theta = 38.66°$$

$v = v_A = 5$ m/s, in the direction shown below:

As v_A is a constant, and $a_t = \dot{v}$

$$a_t = 0$$

∴ a_n is perpendicular to the path of motion.

Now $a_n = v^2 / \rho = 5^2 / 31.5$

$a_n = 0.794\text{ m/s}^2$, in the direction shown.

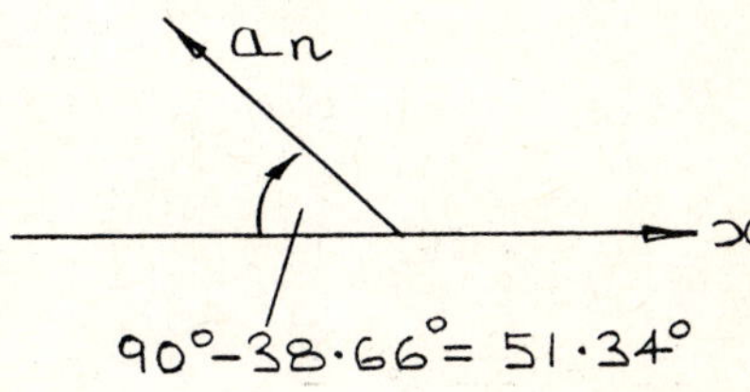

Example 2.9

A car starts from rest at a Point A and accelerates along this horizontal track with a value of $a_t = 0.4\,t$, into the semi-circular bend. Determine the velocity and acceleration at the Point B.

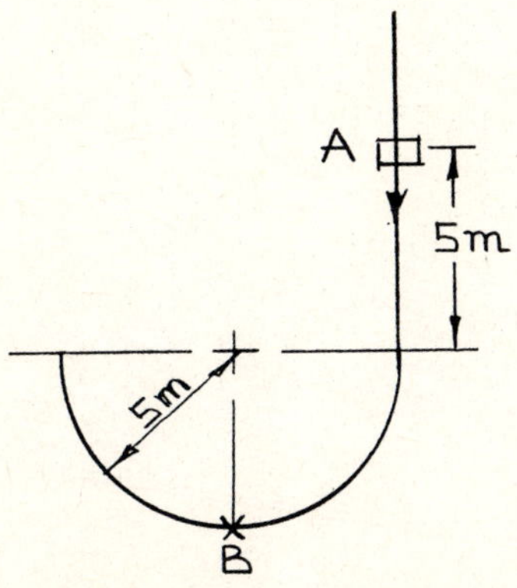

Figure 2.14

Now

$$a_t = \frac{dv}{dt} = 0.4t$$

or

$$\int_o^t 0.4t\ dt = \int_o^v dv$$

$$\therefore \frac{0.4t^2}{2} = v$$

or

$$v = 0.2t^2 \tag{2.58}$$

Distance from A to B =

$$S_{AB} = 5 + 2\pi \text{ x } \frac{5}{4} = 12.854 \text{ m}$$

Now

$$v = \frac{ds}{dt} = 0.2t^2$$

$$\text{or } \int_o^{12.854} ds = \int_o^{t_B} 0.2t^2\ dt$$

$$\text{or } [s]_o^{12.854} = 0.2 \left[\frac{t^3}{3}\right]_o^{t_B}$$

$$\text{or } 12.854 = 0.0667\,t_B^3$$

$$\therefore\ t_B = 5.777\,\text{s} \tag{2.59}$$

where t_B = time taken to travel from A to B

From equation (2.58)

$$v = 0.2t^2$$

or

$$v_B = \text{velocity at B}$$

$$= 0.2 \times 5.777^2$$

$$v_B = 6.675\,\text{m/s} = v_t \text{ at B}$$

$$v_n = 0$$

Now $a_t = 0.4t$

$\therefore\ a_t$ (at B) $= 0.4 \times 5.777$

or $a_t = 2.31\ \text{m/s}^2$

a_n (at B) $= v^2/\rho = 6.75^2/5$

or $a_n = 8.91\ \text{m/s}^2$

$\therefore$ at B

$$a_n = 8.91\,\text{m/s}^2$$

and

$$a_t = 2.31\,\text{m/s}^2$$

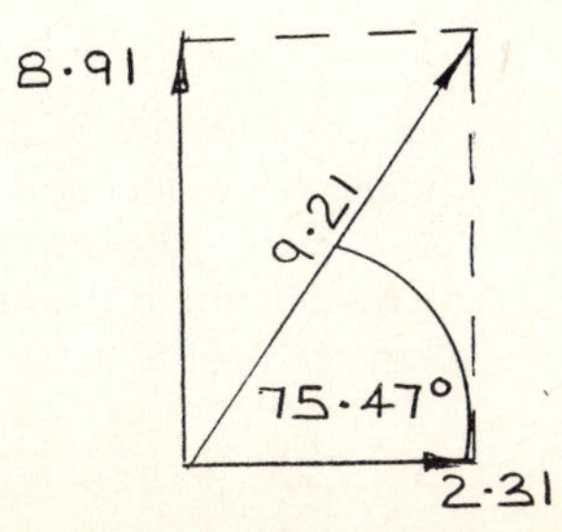

$$a = \sqrt{\left(a_n^2 + a_t^2\right)} = 9.21\,\text{m/s}^2$$

$$\alpha = \tan^{-1}\left(a_n/a_t\right) = 75.47°$$

$\therefore\ a = 9.21\ \text{m/s}^2$, in the direction shown in the sketch.

Problems 2.3

Assume all bodies are particles and that there is no air resistance.

1. The speed of a car is uniformly increased from 40 km per hour to 80 km per hour in 10 seconds, around a circular track of radius 50 m. Determine the magnitude of the acceleration at the end of this 10 second period.

   [9.937 m/s^2]

2. A car travels at a uniform speed of 80 km per hour around a bend with a variable radius. If it takes 10 seconds to travel from the Point 'A' to the Point 'B' and the radii of curvature at these two points are 30 m and 50 m respectively, determine the magnitudes of acceleration at these two points.

   [16.46 m/s^2, 9.87 m/s^2]

3. A ship is turning in a circular path of radius 20 m. If at a certain point in its motion its speed is 10 knots, and its tangential acceleration is 4 m/s^2, determine the magnitude of its acceleration.

   [4.214 m/s^2]

4. A car is travelling at 90 km/h at a Point 'A', when it decelerates uniformly, reaching a speed of 40 km/h at the Point 'C', the distance between A and B and B and C being 50 m, as shown in Figure 2.15. If ρ_A = 10 m and $\bar{a}_c$ = 4 m/s^2, determine the radius of curvature at C and the magnitudes of acceleration at A and B.

   [39.62 m, 12.6 m/s^2, -2.51 m/s^2]

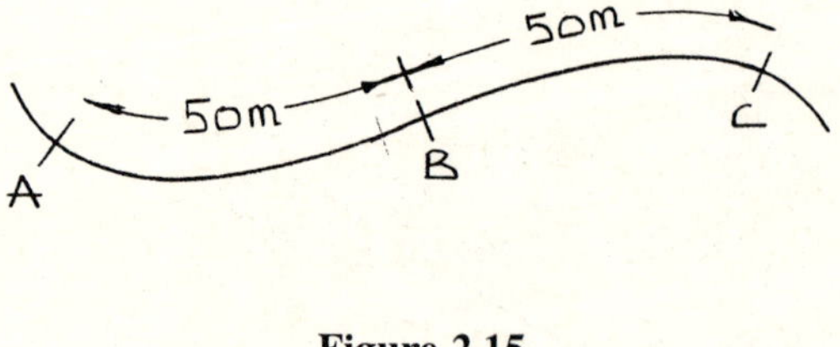

Figure 2.15

5. An aircraft which is flying horizontally, at a uniform speed of 900 km/h, is used to simulate the condition of weightlessness for a short period of time. In order to achieve this, the aircraft must fly in an arc of a circle, as shown in Figure 2.16. Determine the rate of change of α, in degrees per second, at which the pilot has to drop his sight to achieve the desired effect. Take 'g' as 9.8 m/s^2.

   [$\dot{\alpha}$ = 2.25° per second]

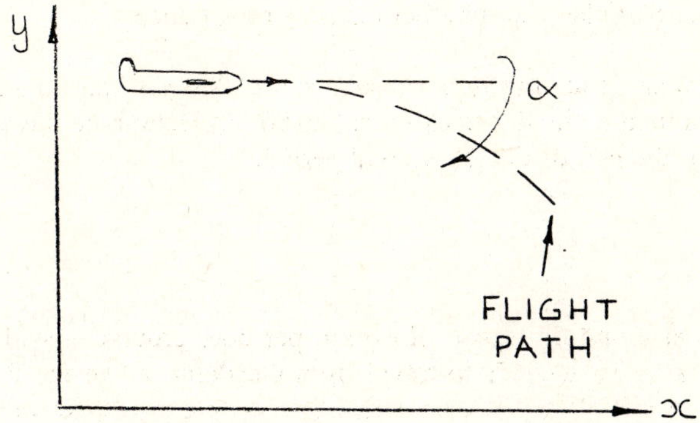

Figure 2.16

2.6 POLAR CO-ORDINATES (r - Θ)

Motion of particles in terms of polar co-ordinates is important, when artefacts rotate about a fixed point. Such cases are many and varied in most branches of engineering, including the motions of mechanisms, etc, in automobile, aircraft and marine engineering.

In this case, the co-ordinates are r and θ, where

r = radius at any time 't'
θ = angle of rotation at any time 't' as shown in Figure 2.17

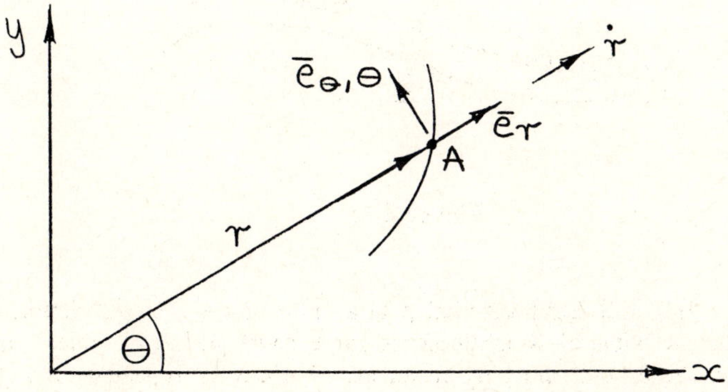

Figure 2.17 Motion of a Particle in Polar Co-ordinates

From Figure 2.17

$\bar{e}_r$ = unit vector in the direction of r

$\bar{e}_\theta$ = unit vector perpendicular to r

$\bar{r}$ = position vector in the direction 'r'

$= r\,\bar{e}_r$

If the particle at the point 'A', moves a small distance along the path of the motion, the vector diagrams for the change of the unit vectors $\bar{e}_r$ and $\bar{e}_\theta$ are shown in Figures 2.18(a) and (b).

Now from Figure 2.18, it can be seen that $d\bar{e}_r$ takes place in the direction $\bar{e}_\theta$ and that $d\bar{e}_\theta$ takes place in the direction $-\bar{e}_r$, so that

$$d\bar{e}_r = \bar{e}_\theta \,.\, d\theta \tag{2.60}$$

$$\text{and } d\bar{e}_\theta = \bar{e}_r \,.\, d\theta \tag{2.61}$$

If we differentiate equations (2.60) and (2.61) with respect to 't', we get

$$\frac{d\bar{e}_r}{dt} = \bar{e}_\theta \,.\, \frac{d\theta}{dt} \tag{2.62}$$

$$\text{and } \frac{d\bar{e}_\theta}{dt} = \bar{e}_r \,.\, \frac{d\theta}{dt} \tag{2.63}$$

$$\text{or } \dot{\bar{e}}_r = \dot{\theta}\,\bar{e}_\theta \tag{2.64}$$

$$\text{and } \dot{\bar{e}}_\theta = -\dot{\theta}\,\bar{e}_r \tag{2.65}$$

$$\text{Now } \bar{r} = r\bar{e}_r$$

$$\text{and } \bar{v} = \frac{d\bar{r}}{dt} = \dot{r}\,\bar{e}_r + r\dot{\bar{e}}_r$$

$$\text{but } \dot{\bar{e}}_r = \dot{\theta}\,\bar{e}_\theta$$

$$\therefore\ \bar{v} = \dot{\bar{r}} = \dot{r}\,\overline{e_r} + r\,\dot{\theta}\,\overline{e_\theta} \tag{2.66}$$

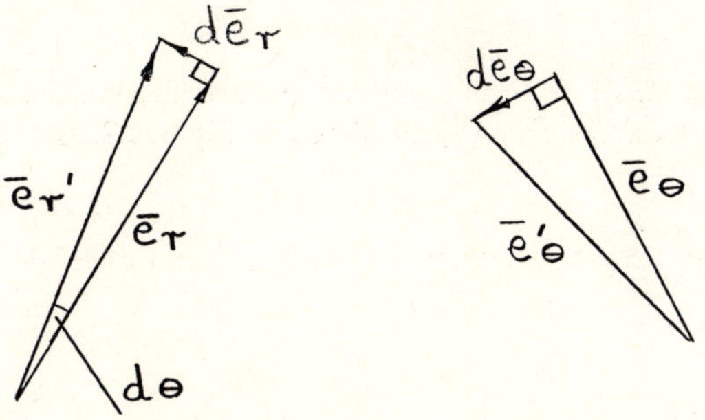

(a) (b)

Figure 2.18 Movement of Unit Vectors

where

$$v_r = \dot{r} = \frac{\mathrm{d}r}{\mathrm{d}t} = \text{radial velocity}$$

$$v_\theta = r\,\dot{\theta} = \text{circumferential velocity}$$

$$v = \sqrt{\left(v_r^2 + v_\theta^2\right)} = \text{resultant velocity} \tag{2.67}$$

which occurs at θ to r.

where $\theta = \tan^{-1}(v_\theta/v_r)$ (2.68)

Similarly,

$$\bar{a} = \frac{\mathrm{d}\,\bar{v}}{\mathrm{d}t}$$

but

$$\overline{v} = \dot{r}\overline{e_r} + r\dot{\theta}\overline{e_\theta}$$

$$\therefore \overline{a} = \ddot{r}\,\overline{e_r} + \dot{r}\overline{\dot{e}_r} + \dot{r}\,\dot{\theta}\,\overline{e_\theta}$$

$$+ r\ddot{\theta}\,\overline{e_\theta} + r\dot{\theta}\,\overline{\dot{e}_\theta}$$

$$= \left(\ddot{r}\overline{e_r} + \dot{r}\overline{\dot{e}_r}\right) + \left(\dot{r}\dot{\theta}\overline{e_\theta} + r\ddot{\theta}\overline{e_\theta} + r\dot{\theta}\overline{\dot{e}_\theta}\right) \tag{2.69}$$

$$\text{but } \overline{\dot{e}_r} = \dot{\theta}\overline{e_\theta}$$

$$\text{and } \overline{\dot{e}_\theta} = -\dot{\theta}\overline{e_r}$$

so that equation (2.69) becomes

$$\overline{a} = \left(\ddot{r}\overline{e_r} + \dot{r}\,.\,\dot{\theta}\overline{e_\theta}\right) + \left(\dot{r}\dot{\theta}\overline{e_\theta} + r\ddot{\theta}\overline{e_\theta} - r\dot{\theta}\,.\,\dot{\theta}\overline{e_r}\right)$$

$$\text{or } \overline{a} = \left(\ddot{r} - r\dot{\theta}^2\right)\overline{e_r} + \left(r\ddot{\theta} + 2\dot{r}\dot{\theta}\right)\overline{e_\theta} \tag{2.70}$$

$$\text{where } a_r = \ddot{r} - r\dot{\theta}^2$$

$$a_\theta = r\ddot{\theta} + 2\dot{r}\dot{\theta} \tag{2.71}$$

$$\text{and } a = \sqrt{\left(a_r^2 + a_\theta^2\right)}$$

$\ddot{r} = \dfrac{\mathrm{d}^2 r}{\mathrm{d}t^2}$ = acceleration due to the rate of change of the rate of change of the radial length r with respect to time t

$-r\dot{\theta}^2$ = centripetal acceleration. This is an inward acceleration caused by the particle rotating at an angular velocity of '$\dot{\theta}$' at a radius 'r'.

$r\ddot{\theta}$ = acceleration caused by the particle suffering an angular acceleration of '$\ddot{\theta}$' at a radius 'r'.

$2\dot{r}\dot{\theta}$ = **Coriolis component** of the angular acceleration.

Explanation of the **Coriolis acceleration** can be described as follows:

If a fairground worker walks radially outwards on a fairground roundabout, at a uniform velocity of $\dot{r}$ and at the same time the roundabout is rotating at an uniform angular velocity of $\dot{\theta}$, the fairground worker will experience a tangential acceleration of $2\dot{r}\dot{\theta}$; this acceleration is called the **Coriolis acceleration**.

Example 2.10

A collar 'B' which may be regarded as a particle, fits on to a uniform section rod, which rotates about a pin 'A', as shown in Figure 2.19. The rod rotates according to the expression $\theta = t^4/2$ and the collar slides outwards according to the expression $r = 90\, t^2$mm, where 't' is measured in seconds. Determine the velocity and acceleration of the collar, when $t = 0.9$ seconds.

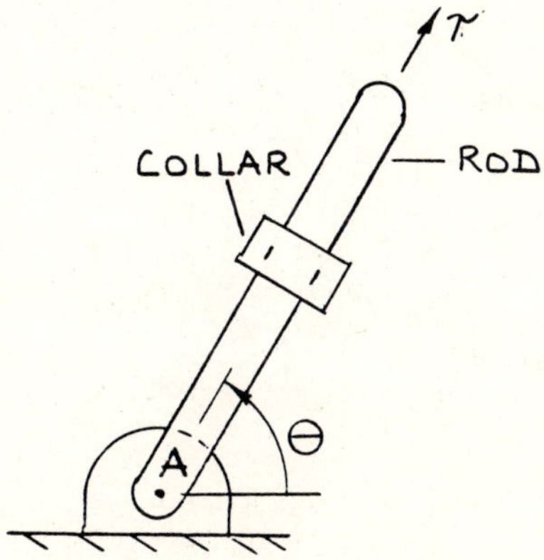

Figure 2.19 Collar Sliding on Rod

Now $\theta = t^4/2$ rads

$\therefore$ when t = 0.9 seconds

θ = 0.328 rads = 18.8°

and r = $90\, t^2$ mm = 72.9 mm

From equation (2.66)

$$\bar{v} = \dot{r}\,\bar{e}_r + r\,\dot{\theta}\bar{e}_\theta$$

$$= 180\,t\,\bar{e}_r + 90\,t^2 \text{ x } 2\,t^3$$

when t = 0.9 seconds

$$\bar{v} = 162\,\bar{e}_r + 106.3\,\bar{e}_\theta \text{ mm/s}$$

or v_r = 162 mm/s

and v_θ = 106.3 mm/s

v (at B) = $\sqrt{(162^2 + 106.3^2)}$ = 193.8 mm/s

$\beta = \tan^{-1}(106.3/162) = 33.28°$

∴ Angle that 'v' acts with the horizontal is $\theta + \beta$

= 18.8 + 33.28 = 52.1° (see the figure below)

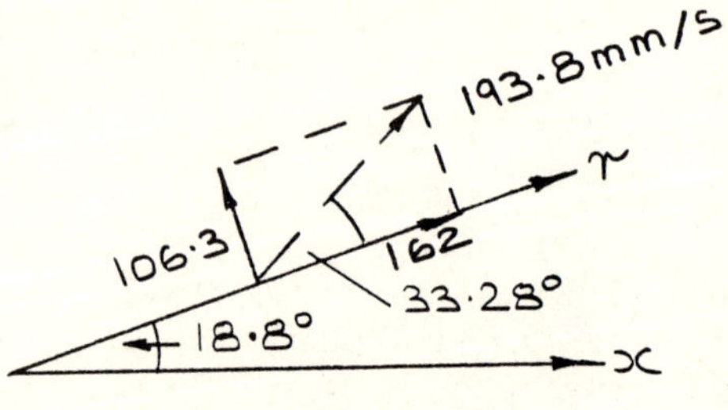

From equation (2.70)

$$\bar{a} = \left(\ddot{r} - r\dot{\theta}^2\right)\bar{e}_r + \left(r\ddot{\theta} + 2\dot{r}\dot{\theta}\right)\bar{e}_\theta$$

$$= \left[180 - 90t^2 \text{ x } \left(2t^3\right)^2\right]\bar{e}_r$$

$$+ \left(90t^2 \text{ x } 6t^2 + 2 \text{ x } 180t \text{ x } 2t^3\right)\bar{e}_\theta$$

$$= (180 - 236.2)\,\overline{e_r} + (354.3 + 472.4)\,\overline{e_\theta}$$

or $\overline{a} = -56.2\overline{e_r} + 826.7\overline{e_\theta}$

$a_r = -56.2$ mm/s^2 and $a_\theta = 862.7$ mm/s^2

$a = \sqrt{(56.2^2 + 826.7^2)} = 828.6$ mm/s^2

$\alpha = \tan^{-1}(-826.7/56.2) = 86.1°$

(See the figure below)

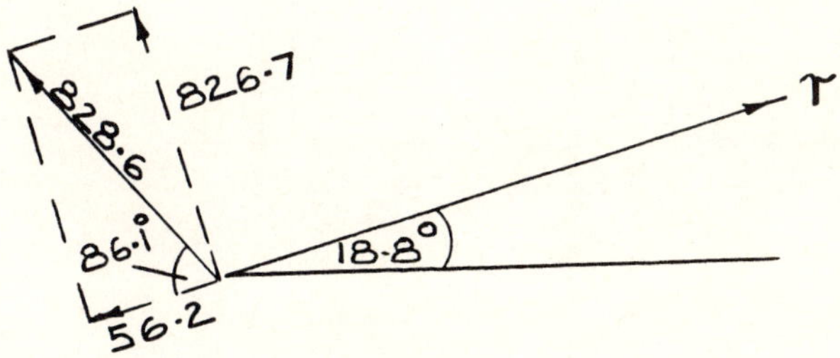

Problem 2.4

Regard all the relevant objects as particles.

1. A collar B fits snugly onto a uniform section rod, which rotates about the pin A, as shown in Figure 2.19. If the rod rotates according to the expression $\theta = 1 + t^2$mm and the collar slides outwards according to the expression $r = 0.5 + 2t^2$ mm, determine the velocity and acceleration of the collar when $t = 1.5$ seconds.

   [15.13 mm/s, $\alpha = 75.1°$, 46.53 mm/s^2, $\beta = 28.22°$]

2. A particle has plane curvilinear motion according to the expressions:

   $r = r_o\, e^{\alpha t}$ and $\theta = \beta t$.

   Determine expressions for the velocity and acceleration of the particle at the time 't'.

   [$v = r_o\, e^{\alpha t} \sqrt{(\alpha^2 + \beta^2)}$, $a = r_o\, e^{\alpha t} (\alpha^2 + \beta^2)$]

3. A searchlight which traverses a wall in a vertical plane, casts a spot of light on the wall as the searchlight rotates about the point '0'. The angular velocity of the spot is $\dot{\theta}$ = 5 rads/s, as shown in Figure 2.20. If the searchlight is 80 m from the wall, determine the velocity and acceleration of the spot of light when θ = 30°.

[533.3 m/s, 6711.1 m/s^2]

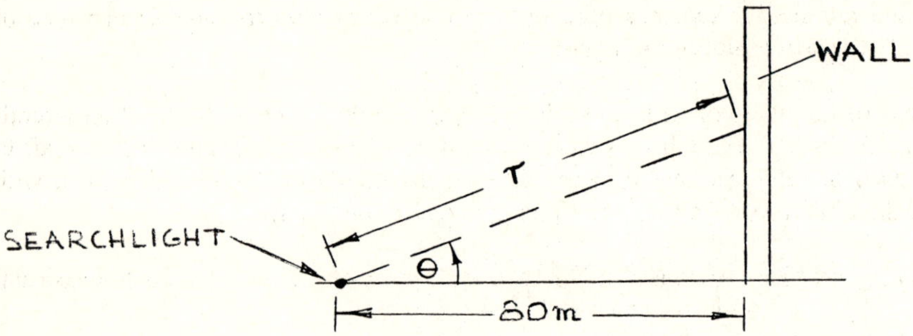

Figure 2.20

4. A car is travelling along level ground on a circular bend of radius 200 m, at a rate of angular velocity of $\dot{\theta}$ = 0.04 rads/s and an angular acceleration of $\ddot{\theta}$ = 0.004 rads/s^2. Determine the velocity and acceleration of the car.

[8 m/s, 0.86 m/s^2]

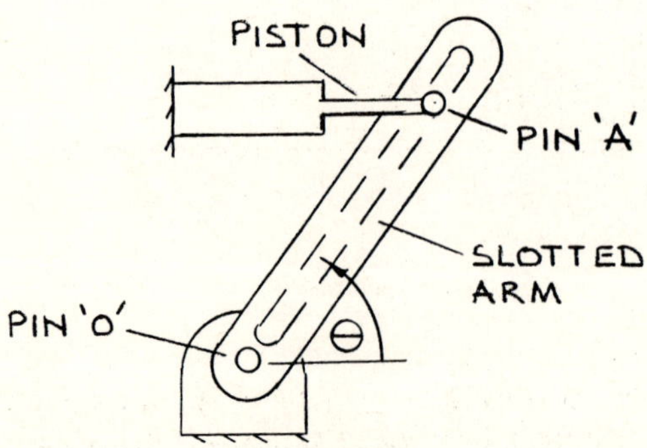

Figure 2.21

5. A mechanism is such that a horizontal piston exerts a uniform horizontal velocity of 0.8 m/s to a pin A on a slotted arm, free to rotate about the pin '0'. When $\theta = 75°$, determine $\dot{r}$, $\ddot{r}$, $\dot{\theta}$ and $\ddot{\theta}$.

[-207.1 mm/s, 2881.4 mm/s^2, 3.73 rads/s, 7.46 rads/s^2]

2.7 RELATIVE MOTION USING TRANSLATING AXES

So far in this chapter, we have considered only the absolute motion of a particle with reference to fixed axes. There are, however, cases where the axes move and it is necessary to consider the relative motion of a particle from one set of axes to another; this type of motion is called **relative motion analysis**.

A typical case of relative motion analysis is that of a rotating propeller blade of an aircraft in flight. In this case, it is usually found convenient to refer to the motion of the aircraft to some fixed axes of reference and then to consider the motion of the propeller blade with reference to the aircraft, where the latter motion is added vectorially.

Prior to solving problems in this field, it will be necessary to make some fundamental definitions.

2.7.1 Positions

Consider the separate motion of two particles, namely particle A and particle B, who follow the independent paths aa and bb respectively, as shown in Figure 2.22.

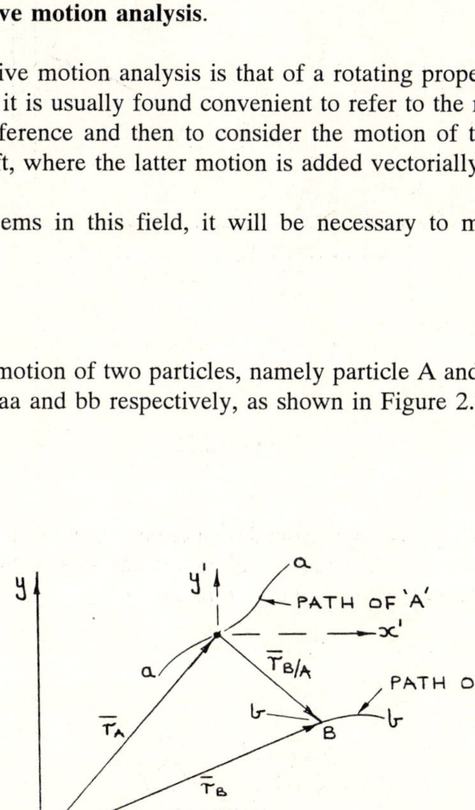

Figure 2.22

The absolute positions of the two particles 'A' and 'B' from the x-y origin are $\bar{r}_A$ and $\bar{r}_B$. As the path of 'B' rotates and translates relative to the path of 'A', it will be convenient to use another set of rectangular axes, namely the x^1-y^1 axes; these axes translate from the x-y axes but do not rotate. Hence, the relative position of 'B' with reference to 'A' is given by the vector $\bar{r}_{B/A}$, so that

$$\bar{r}_B = \bar{r}_A + \bar{r}_{B/A} \tag{2.72}$$

where $\bar{r}_{B/A}$ is the position of 'B' relative to 'A'.

2.7.2 Velocity

In a manner similar to that described above, it can be shown that the velocity vectors of the particle are related by the expression:

$$\bar{v}_B = \bar{v}_A + \bar{v}_{B/A} \tag{2.73}$$

as shown in Figure 2.23.

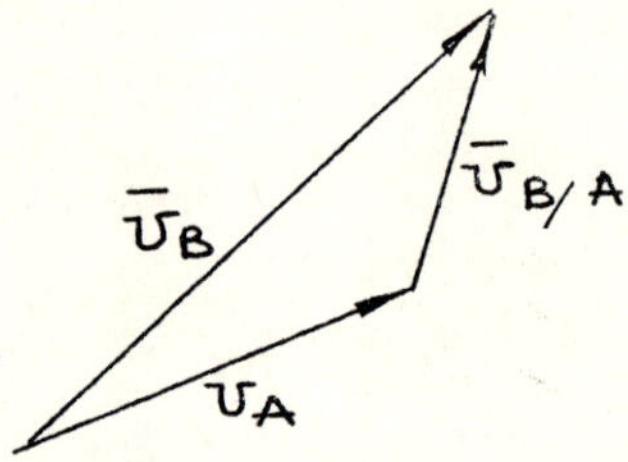

Figure 2.23 Velocity Vectors

The velocity vectors, namely, $\bar{v}_A$, $\bar{v}_B$ and $\bar{v}_{B/A}$ can be related to the position vectors, namely $\bar{r}_A$, $\bar{r}_B$ and $\bar{r}_{B/A}$ as follows:

$$\begin{aligned} \overline{v_A} &= \frac{d\overline{r_A}}{dt} \\ \overline{v_B} &= \frac{d\overline{r_B}}{dt} \\ \overline{v_{B/A}} &= \frac{d\overline{r_{B/A}}}{dt} \end{aligned} \tag{2.74}$$

where $\bar{v}_A$ and $\bar{v}_B$ are absolute velocity vectors of 'A' and 'B', because they are referred to the fixed x-y axes. The vector $\bar{v}_{B/A}$ is a relative velocity of 'B' with reference to 'A', because it is referred to the moving x^1-y^1 axes of Figure 2.22.

2.7.3 Acceleration

In this case the vectors of acceleration are related by the expression:

$$\overline{a_B} = \overline{a_A} + \overline{a_{B/A}} \tag{2.75}$$

as shown in Figure 2.24.

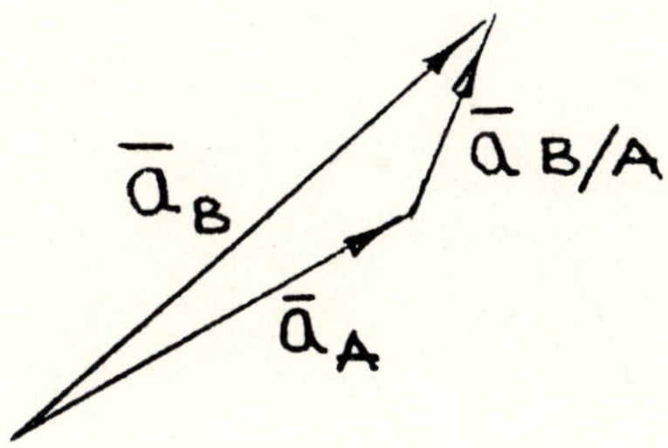

Figure 2.24 Acceleration Vectors

The acceleration vectors are related to the velocity vectors, as follows:

$$\overline{a_A} = \frac{d\overline{v_A}}{dt}$$

$$\overline{a_B} = \frac{d\overline{v_B}}{dt} \tag{2.76}$$

$$\overline{a_{B/A}} = \frac{d\overline{v_{B/A}}}{dt}$$

where $\bar{a}_A$ and $\bar{a}_B$ are absolute accelerations related to the x-y axes and $\bar{a}_{B/A}$ is a relative acceleration, related to the moving x^1-y^1 axes.

Example 2.11

A train track passes over a roadway, as shown in Figure 2.25. If the train is travelling eastwards at 100 km per hour and the car is travelling in the direction shown at 80 km per hour, what is the relative velocity of the train with respect to the car.

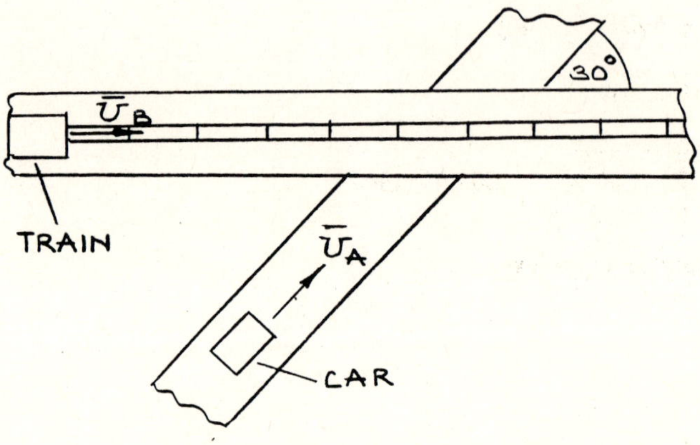

Figure 2.25

The vector diagram for the velocities is shown in Figure 2.26. The vector '$\bar{v}_B$' is drawn to scale in a horizontal direction from a point 'O'. Similarly, the vector '$\bar{v}_A$' is also drawn to scale in the direction in which this velocity acts; also from the point 'O'. The ends of these two vectors are now jointed together, to give the velocity vector $\bar{v}_{B/A}$, as shown in Figure 2.26, where $\bar{v}_{B/A}$ is the velocity of 'B' relative to 'A'.

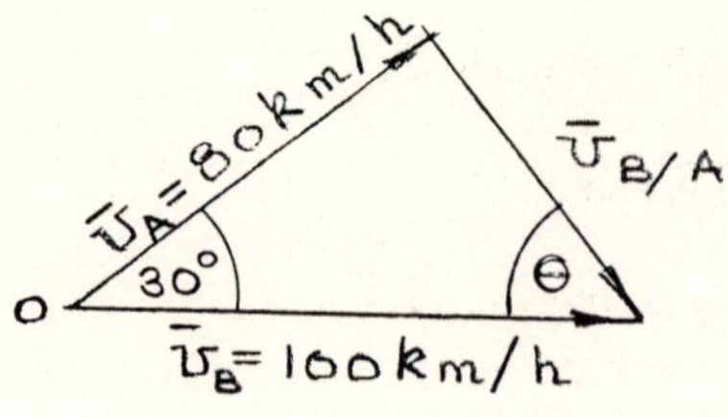

Figure 2.26

From equation (2.73)

$$\bar{v}_B = \bar{v}_A + \bar{v}_{B/A}$$

or in vector analysis

$$100\,\bar{i} = \left(v_A \cos 30\,\bar{i} + v_A \sin 30\,\bar{j}\right) + \overline{v_{B/A}}$$

$$= \left(69.28\bar{i} + 40\bar{j}\right) + \overline{v_{B/A}}$$

$$\text{or } \overline{v_{B/A}} = \left(100 - 69.28\right)\bar{i} - 40\bar{j}$$

$$= 30.72\bar{i} - 40\bar{j}$$

$$= \sqrt{\left(30.72^2 + 40^2\right)} = 50.43 \text{ km/h}$$

The angle 'θ' is found from the expression:

$$\tan\theta = \frac{\left(v_{B/A}\right)_y}{\left(v_{B/A}\right)_x}$$

where

$$(v_{B/A})_x = \text{'}x\text{' component of } \bar{v}_{B/A}$$

$$= 30.72 \text{ km per hour}$$

$$(v_{B/A})_y = \text{'}y\text{' component of } \bar{v}_{B/A}$$

$$= -40 \text{ km per hour}$$

$$\therefore \theta = \tan^{-1}(40 / 30.72) = \tan^{-1}(1.302)$$

$$\theta = 52.47°$$

Alternative method of calculating $v_{B/A}$

From Figure 2.26 and the cosine rule of equation (1.7)

$$\overline{v_{B/A}} = \overline{v_A^2} + \overline{v_B^2} - 2 \times \overline{v_A} \times \overline{v_B} \cos 30$$

$$= 80^2 + 100^2 - 2 \times 80 \times 100 \cos 30$$

$$= 16400 - 13856$$

$$\text{or } \overline{v_{B/A}} = 50.43 \text{ km per hour}$$

From the sine rule of equation (1.6)

$$\frac{\overline{v_{B/A}}}{\sin 30} = \frac{\overline{v_A}}{\sin\theta}$$

or $$\frac{50.43}{\sin 30} = \frac{80}{\sin\theta}$$

$$\therefore \sin\theta = 80 \text{ x } \sin 30 \,/\, 50.43$$

$$= 0.793$$

$$\theta = \sin^{-1}(0.793)$$

$$\theta = 52.47°$$

Now $\overline{v}_{B/A} = 50.43 \cos(52.47)\,\overline{i} - 50.43 \sin(52.47)\,\overline{j}$

$\overline{v}_{B/A} = 30.72\,\overline{i} - 40\,\overline{j}$ (as before)

Example 2.12

A plane is travelling westwards at a speed of 850 km per hour, when another plane, namely plane 'B' flies under it in the direction shown. If plane 'B' appears to be travelling in a direction of 60° relative to the direction of plane 'A', determine the velocity of plane 'B'. What is the relative velocity of plane 'B' to plane 'A'.

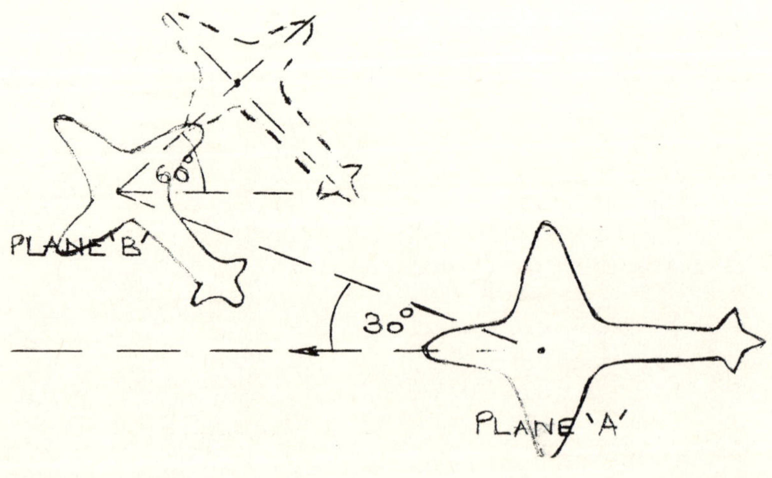

Figure 2.27

The velocity diagram is drawn in Figure 2.28, as follows. From the point 'O', draw in the velocity of plane 'A' to scale, from the right of the page to the left. Now the velocity of 'B' is not known, but its direction is, so draw a line from 'O' at 30°, as shown in Figure 2.28. Next, from the point 'a' in Figure 2.28, draw a line in the direction shown to represent the velocity vector of 'B' with reference to 'A', as shown. These two lines intersect at 'b', hence, $\overline{v_B}$ and $\overline{v_{B/A}}$ can be determined.

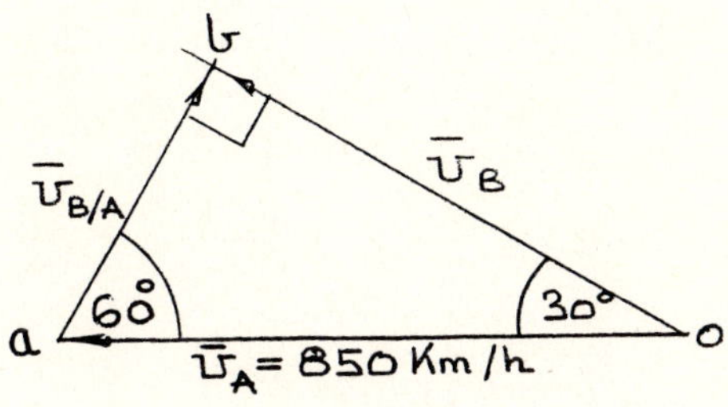

Figure 2.28

From Figure 2.28 and the sine rule of equation (1.6)

$$\frac{\overline{v_B}}{\sin 60} = \frac{850}{\sin 90}$$

$$\overline{v_B} = 850 \times 0.866$$

$$\overline{v_B} = 736.1 \text{ km per hour}$$

Again from the sine rule,

$$\frac{\overline{v_{B/A}}}{\sin 30} = \frac{850}{\sin 90}$$

$$\therefore \overline{v_{B/A}} = 850 \times 0.5 = 425 \text{ km per hour}$$

Problem 2.5

1. Two boats, namely boat 'A' and boat 'B' leave the shore at the same time but travel in different directions, as shown in Figure 2.29. Determine the velocity of 'B' relative to 'A'.

   $[v_{B/A} = \sqrt{(v_A^2 + v_B^2 - 2v_A\, v_B \cos\theta)}]$

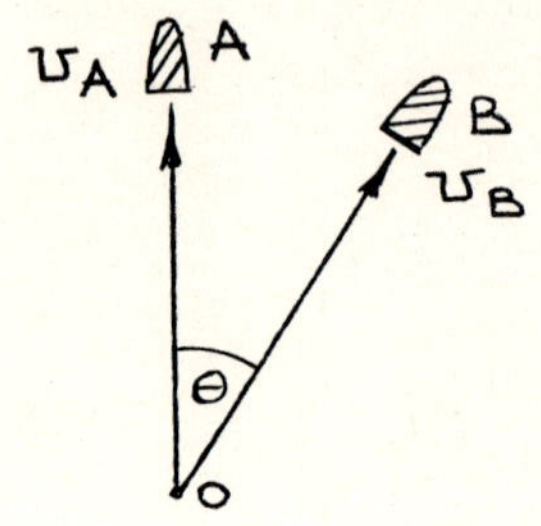

Figure 2.29

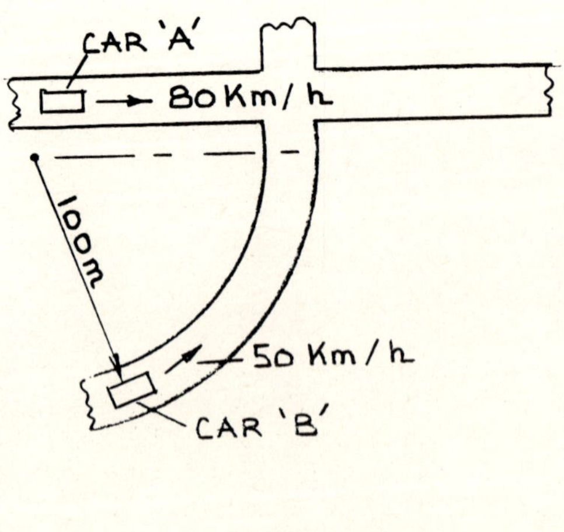

Figure 2.30

2. Car ‘A’ is accelerating in an eastwards direction at 1.5 m/s^2, while car ‘B’ is rounding a bend of radius 100 m at a uniform speed of 50 km per hour, as shown in Figure 2.30. If at a certain time, car ‘A’ is travelling at 80 km per hour, determine the relative velocity and acceleration that car ‘B’ appears to have to a passenger in car ‘A’.

[44.41 km per hour, 34.26°]

3. Planes ‘A’ and ‘B’ are travelling at different altitudes in the directions shown in Figure 2.31. If plane ‘A’ is travelling at 800 km per hour and plane ‘B’ is travelling at 600 km per hour, at what velocity will plane ‘B’ appearing to be travelling to a passenger in plane ‘A’.

[1353 km per hour, 12.81°]

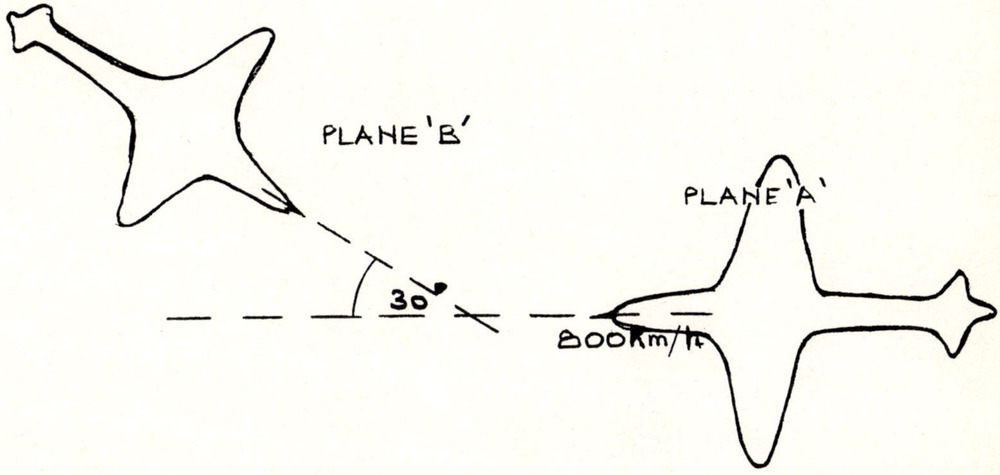

Figure 2.31

4. A certain time, two cars, namely car ‘A’ and car ‘B’ are travelling at speeds of 100 km per hour and 60 km per hour, respectively, as shown in Figure 2.32. If the speed of ‘A’ is constant, but ‘B’ is increasing its speed at 2.5 m/s^2, determine the velocity and acceleration of ‘B’ relative to ‘A’. It may be assumed that car ‘B’ is travelling around a curve of radius 100m.

[14 m/s, 52.5°, 5.33 m/s^2, 93.14°]

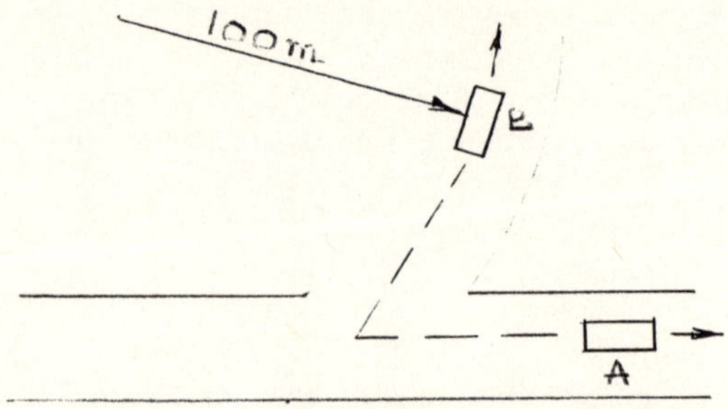

Figure 2.32

5. A slider-crank mechanism is shown in Figure 2.33, where the crank is of 0.1 m radius. The crank rotates in a clockwise direction, about a fixed point 'C' with an angular velocity of 70 rads/s and an angular acceleration of 1000 rads/s^2. The slider moves horizontally in a smooth frictionless guide, where the connecting rod, (or conrod), is the link AB. If the centre of gravity of the conrod is at 'G', determine the velocity and acceleration of 'G' and the angular velocity and angular acceleration of the conrod.

[7.15 m/s, 377 m/s^2, 23.38 rads/s, 2000 rads/s^2]

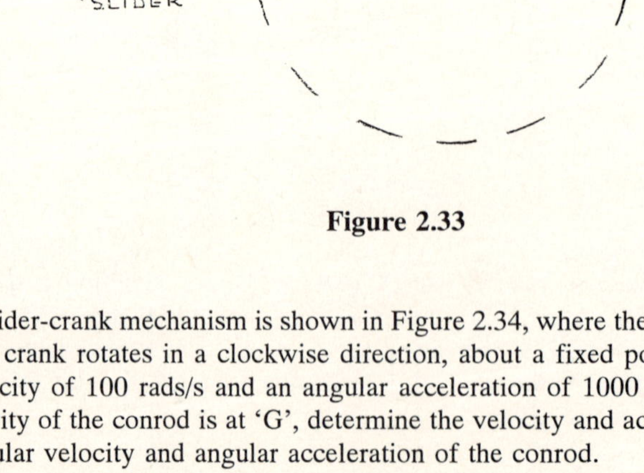

Figure 2.33

6. A slider-crank mechanism is shown in Figure 2.34, where the crank is of 0.1 m radius. The crank rotates in a clockwise direction, about a fixed point 'C', with an angular velocity of 100 rads/s and an angular acceleration of 1000 rads/s^2. If the centre of gravity of the conrod is at 'G', determine the velocity and acceleration of 'G' and the angular velocity and angular acceleration of the conrod.

[8.4 m/s, 600 m/s^2, 33.31 rads/s, 5565 rads/s^2]

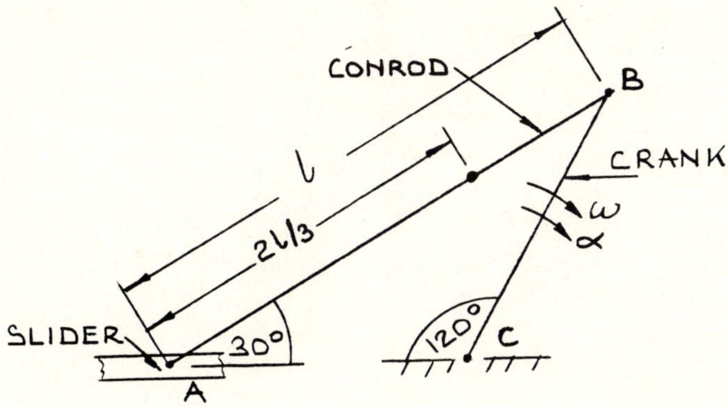

Figure 2.34

2.8 CONSTRAINED MOTION OF CONNECTED PARTICLES

In engineering there are a number of occasions when it is required to understand the motion of interconnected particles. A typical case of this situation is shown in Figure 2.35.

In Figure 2.35, it can be seen that the total length of the cord 'L' is given by:

$$L = x + 2y + \pi r_1 + \pi r_2/2 + c \qquad (2.77)$$

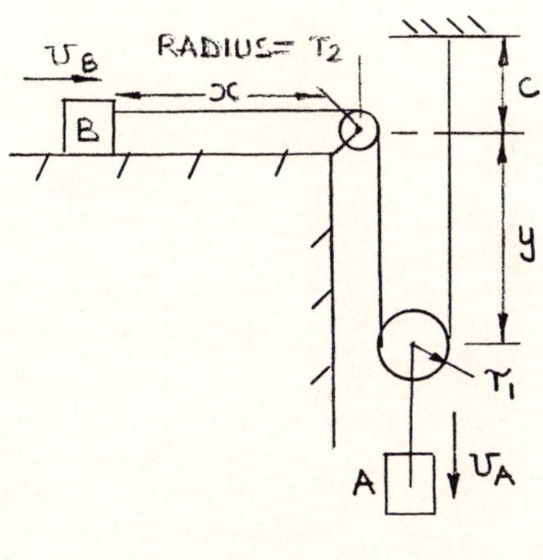

Figure 2.35

If we differentiate equation (2.77) with respect to 't', we get:

$$\frac{dL}{dt} = o = \dot{x} + 2\dot{y} \tag{2.78}$$

or $\dot{x} = -2\dot{y}$

but $\dot{x} = -v_B$, because 'x' decreases its value in the positive direction of v_B

and $\dot{y} = v_A$

$$\therefore v_B = 2v_A \tag{2.79}$$

ie the particle 'B' will travel twice as fast as the particle 'A'.

Similarly, if we differentiate equation (2.78) with respect to 't', we get,

$$\ddot{x} + 2\ddot{y} = 0 \tag{2.80}$$

but $\ddot{x} = -a_B$

and $\ddot{y} = a_A$

$$\therefore a_B = 2a_A \tag{2.81}$$

ie the acceleration of 'B' is twice the value of the acceleration of 'A'.

The revelations of equations (2.79) and (2.81) show us that through the use of a suitable system of pulleys, it should be possible to move a large weight with a small force.

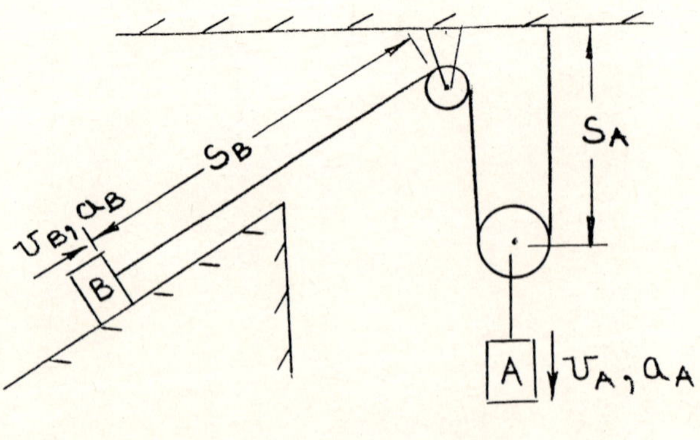

Figure 2.36

Example 2.13

If the acceleration of the block 'B' up the slope is 0.1 m/s^2, determine the downwards acceleration of 'A'.

Now the total length of the cord

$$L = 2\,s_A + s_B + c \qquad (2.82)$$

Differentiating equation (2.82) with respect to 't', we get

$$\frac{dL}{dt} = 0 = 2\dot{s}_A + \dot{s}_B \qquad (2.83)$$

Differentiating equation (2.83) with respect to 't', we get

$$0 = 2\,\ddot{s}_A + \ddot{s}_B \qquad (2.84)$$

but $a_B = -\ddot{s}_B$, because s_B is shortening in the positive direction of a_B

and $a_A = \ddot{s}_A$

$\therefore a_A = a_B/2$

$a_A = 0.05$ m/s^2

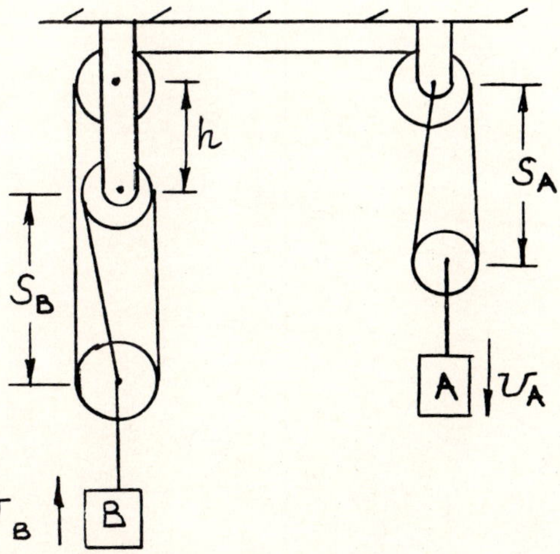

Figure 2.37

Example 2.14

If in the pulley system shown in Figure 2.37, particle 'A' moves downwards with a velocity of 0.5 m/s, what will be the upward velocity of particle 'B'.

Total length of cord

$$L = 3\,s_B + 2\,s_A + h + c \tag{2.85}$$

$$\text{or } \frac{dL}{dt} = o = 3\dot{s}_B + 2\dot{s}_A$$

Now

$v_B = -\dot{s}_B$, because s_B shortens with positive v_B

and $v_A = \dot{s}_A$

$$\therefore v_B = 2v_A/3$$

$$= 2 \text{ x } 0.5/3$$

$$v_B = 0.333 \text{ m/s}$$

Problems 2.6

1. Determine the relationships between v_A and v_B, and a_A and a_B, for the interconnected particles of Figure 2.38.

   [$v_A = v_B$, $a_A = a_B$]

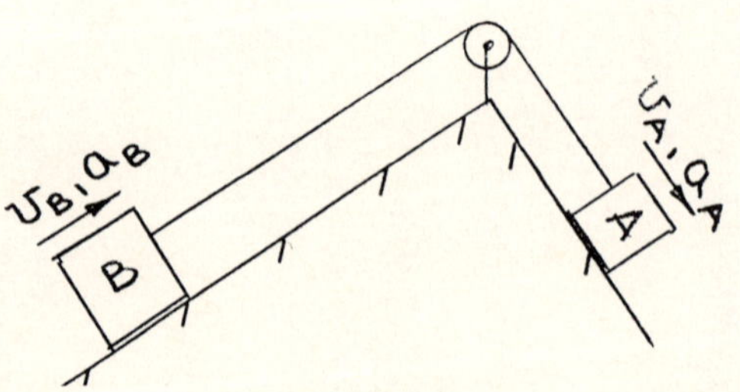

Figure 2.38

2. Determine the relationship between v_B and v_A for the system of interconnected particles shown in Figure 2.39.

[$v_B = v_A/2$]

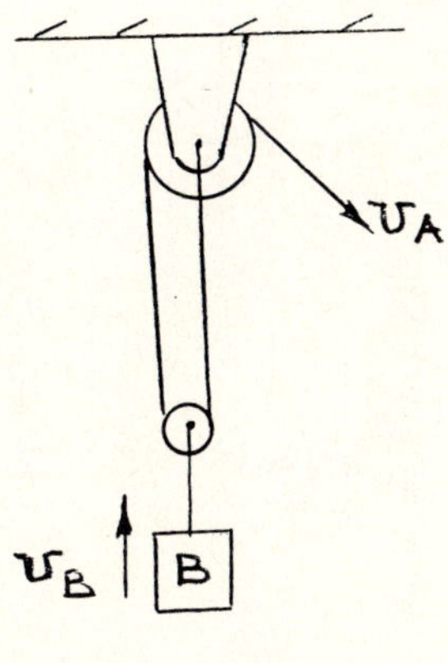

Figure 2.39

3. Determine the values for v_B and a_B, for the system of interconnected particles shown in Figure 2.40.

[4 m/s, 0.2 m/s^2]

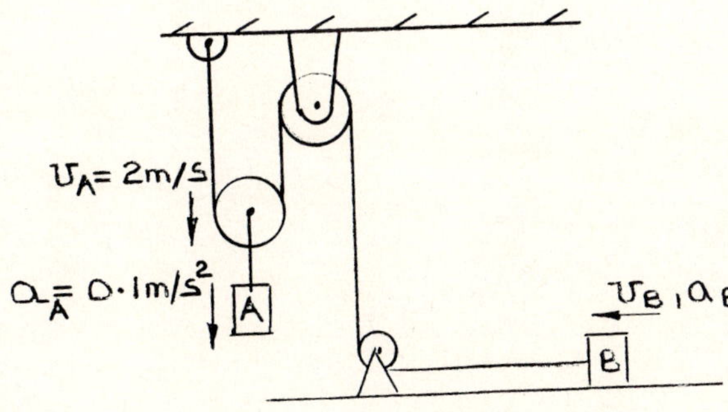

Figure 2.40

4. Given that v_B = 12 m/s, determine v_A for the system of interconnected particles shown in Figure 2.41.

[4 m/s]

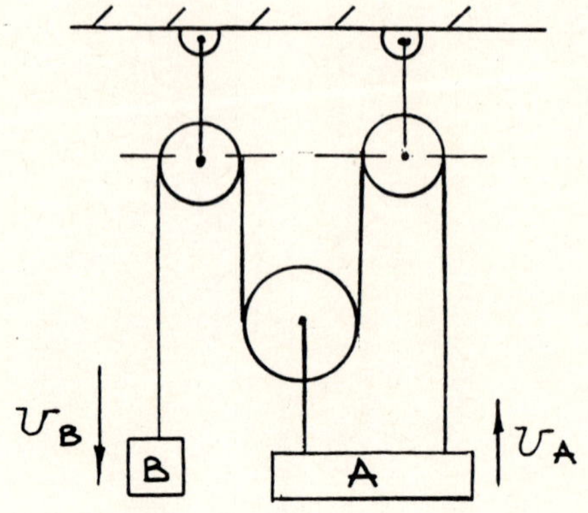

Figure 2.41

5. The motion of B in the system of interconnected particles of Figure 2.42, is given by the expression: $y = 0.5\ t^2$. Determine the acceleration of particle 'A'.

[16 m/s^2]

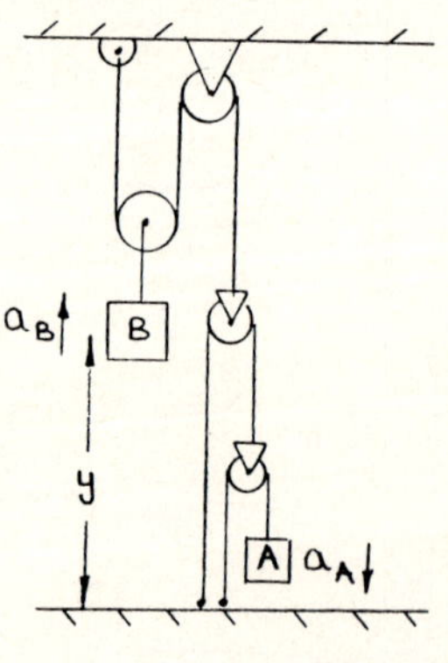

Figure 2.42

6. Determine the three simultaneous equations which describe the motions of the particles A, B, C and D of Figure 2.43.

[$\dot{y}_A + 2\dot{y}_1 = 0$, $2\dot{y}_B - \dot{y}_1 + 2\dot{y}_2 = 0$, $2\dot{y}_c + \dot{y}_D - \dot{y}_2 = 0$]

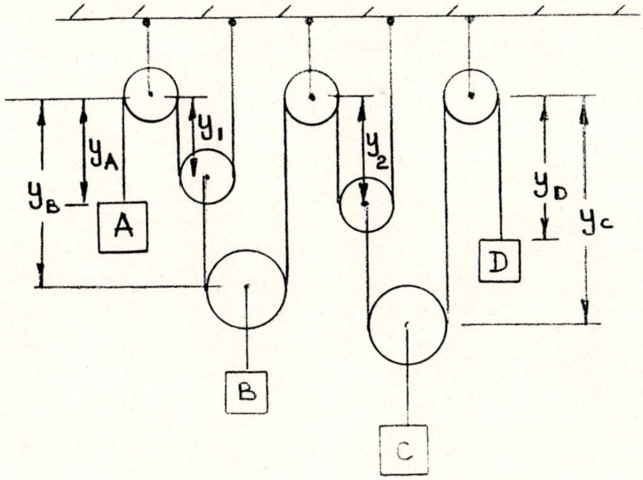

Figure 2.43

CHAPTER THREE

Kinetics of Particles: Force, Mass and Acceleration

3.1 INTRODUCTION

In this chapter, Newton's second law of motion will be applied to problems in dynamics. That is, the relationship between the unbalanced force 'F', the mass 'm' and the acceleration '$\bar{a}$' will be applied to a number of problems in engineering. Both rectilinear motion and motion in rectangular co-ordinates will be considered, together with the motion of particles in normal and tangential co-ordinates and also polar co-ordinates.

3.2 APPLICATION OF NEWTON'S 2ND LAW OF MOTION

From Chapter 1, Newton's 2nd Law of Motion states that:

$$F = \frac{\mathrm{d}(m\,\bar{v})}{\mathrm{d}t}$$

and if 'm' is a constant

$$F = m\,\frac{\mathrm{d}\bar{v}}{\mathrm{d}t} = m\bar{a} \tag{3.1}$$

where F = the unbalanced or resultant force
m = the mass
$\bar{a}$ = the acceleration in the direction of 'F'
$\bar{v}$ = the velocity
t = time

Equation (3.1) can be explained with the aid of Figure 3.1.

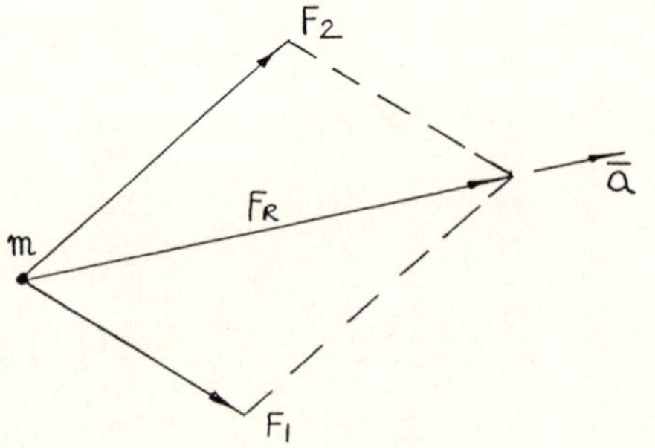

Figure 3.1

Consider the particle of mass 'm' in Figure 3.1. This particle is subjected to two forces 'F_1' and 'F_2' and the resultant of these forces is 'F_R', which can be obtained with the aid of a parallelogram of forces. Applying Newton's 2nd law of motion to this system, it can be seen that the resulting acceleration '$\bar{a}$' is in the direction of F_R, according to the expression:

$$F_R = m\,\bar{a} \tag{3.2}$$

In general, if a mass of value 'm' is subjected to a resultant force 'F_1' it will accelerate at a_1, so that,

$$m = \frac{F_1}{a_1} \tag{3.3}$$

Similarly, if the same mass were subjected to an resultant force 'F_2', it will accelerate with a value of acceleration of a_2, so that, in general,

$$m = \frac{F_1}{a_1} = \frac{F_2}{a_2} = \frac{F_3}{a_3} = \text{etc} = \text{C} \tag{3.4}$$

where, C = a constant

= the inertia of the system

3.3 EQUATIONS OF MOTION IN RECTANGULAR CO-ORDINATES

In a three dimensional rectangular co-ordinate system, Newton's 2nd law of motion becomes:

$$\sum F_x + \sum F_y + \sum F_z$$

$$= m\left(a_x \bar{i} + a_y \bar{j} + a_z \bar{k}\right) \tag{3.5}$$

where

ΣF_x = $m.a_x\,\bar{i}$ = the unbalanced force in the 'x' direction

ΣF_y = $m.a_y\,\bar{j}$ = the unbalanced force in the 'y' direction

ΣF_z = $m.a_z\,\bar{k}$ = the unbalanced force in the 'z' direction

$\bar{i}, \bar{j}$ & $\bar{k}$ = unit vectors in the x, y and z directions, respectively

In **rectilinear motion** equation (3.5) becomes

$$\bar{F} = \sum F_x = m\,.\,a_x \tag{3.6}$$

Example 3.1

A crate of mass 40 kg is subjected to the force shown in Figure 3.2. If the co-efficient of kinetic friction, namely μ_k is 0.25, determine the velocity of the crate after 3s of motion, starting from rest.

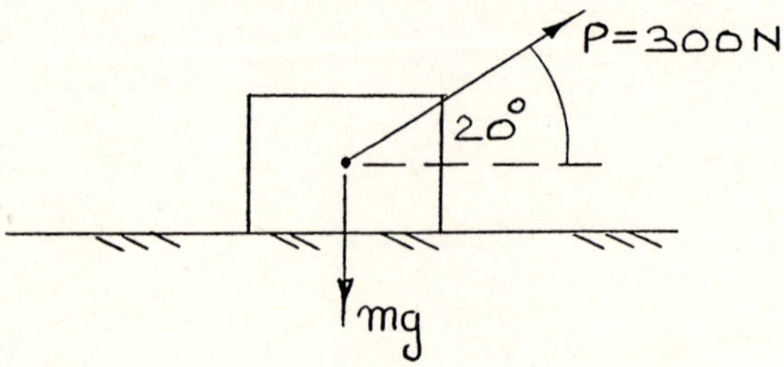

Figure 3.2

Horizontal component of $P = P_x = P \cos 20°$

$= 300 \cos 20°$

$= \underline{281.9 \text{ N}}$

Vertical component of $P = P_y = P \sin 20°$

$= 300 \sin 20°$

$= \underline{102.6 \text{ N}}$

Weight of mass $= mg$

The free body diagram is shown in Figure 3.3.

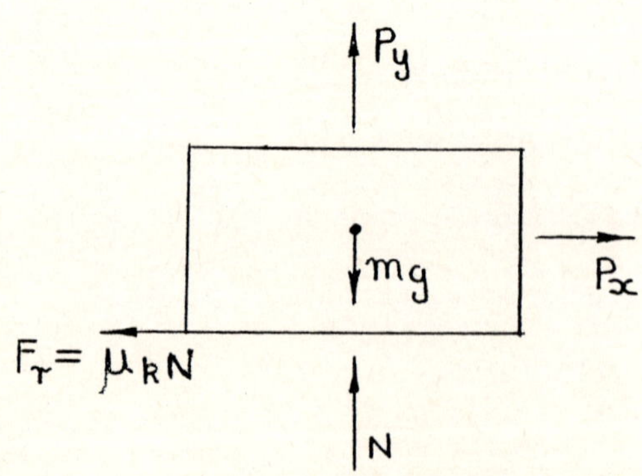

Figure 3.3 Free Body Diagram

Resolving Vertically

Upward forces = Downward forces

or $N + P_y = mg$

$\therefore N = mg - P_y$

$= 40 \times 9.81 - 102.6$

$\underline{N = 289.8 \text{ N}}$

Now the frictional force =

$F_r = \mu_k N = 0.25 \times 289.8$

or $\underline{F_r = 72.45 \text{ N}}$

Now from Newton's 2nd law of motion,

$F = ma =$ resultant force

but from Figure 3.3

$F = P_x - F_r = 281.9 - 72.45$

$\therefore F = 209.45$ N

Hence,

$209.45 = 40\,a$

or $a = 5.236 \text{ m/s}^2$

In this case 'a' is a constant, hence, equation (2.8) can be used to determine 'v'

ie, $v = v_o + at$

$= 0 + 5.236 \dfrac{m}{s^2} \times 3s$

or $v =$ velocity after 3s

$= 15.708$ m/s

Example 3.2

The model of a ship is towed through water in a experimental towing tank, where its resistance to motion, namely 'R', was found to be equal to $1.5\,v^2$ N. If the mass of this model is 15 kg, and the model is released from the towing apparatus when its speed reaches

4 m/s, determine the distance that the model will travel before its speed slows down to 2 m/s. Additionally, over what time does this motion take place?

(a) $$R = -1.5\, v^2 = ma$$

$$\text{but } a = v \frac{dv}{ds}$$

$$\therefore -1.5\, v^2 = mv \cdot \frac{dv}{ds}$$

$$\text{or } -\int_o^s ds = \frac{m}{1.5} \int_{v_o}^{v} \frac{dv}{v}$$

$$\text{or } -s = \frac{m}{1.5} \left[\ln v\right]_{v_o}^{v}$$

$$= \frac{15}{1.5} \left[\ln(2) - \ln(4)\right]$$

$$= 10\,(0.6931 - 1.3863)$$

$$\text{or } s = 6.932\,\text{m}$$

(b) $$\text{Now, } F = ma$$

$$\text{or } -1.5\, v^2 = m \frac{dv}{dt}$$

$$\text{or } \int_o^t -dt = \frac{m}{1.5} \int_{v_o}^{v} \frac{dv}{v^2} = \frac{-m}{1.5} \left[\frac{1}{v}\right]_{v_o}^{v}$$

$$\text{or } -t = \frac{-m}{1.5} \left[\frac{1}{v} - \frac{1}{v_o}\right]$$

$$\text{or } t = 10 \left[\frac{1}{2} - \frac{1}{4}\right]$$

$$t = 2.5\,\text{s}$$

Example 3.3

A mechanism consists of a vertical shaft on which slides an annular collar. The annular collar is fitted with a pin, which is free to rotate in a vertical plane, as shown in Figure 3.4. The pin is subjected to a constant force 'F' and the angle that the pin and the force 'F' makes with the vertical shaft, namely 'θ', is given by

$\theta \;=\; kt,$

where k $=$ a constant

When the collar is at rest, $t = \theta = 0$, and when $\theta = \pi/3$, $v = 0$,

where v $=$ the velocity of the collar

If the μ_k is the coefficient of kinetic friction between the collar and the shaft, determine the value of 'F'.

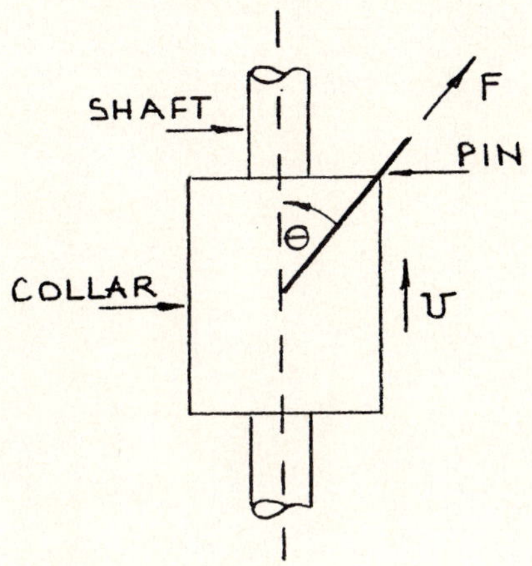

Figure 3.4

The free body diagram of the collar is shown in Figure 3.5

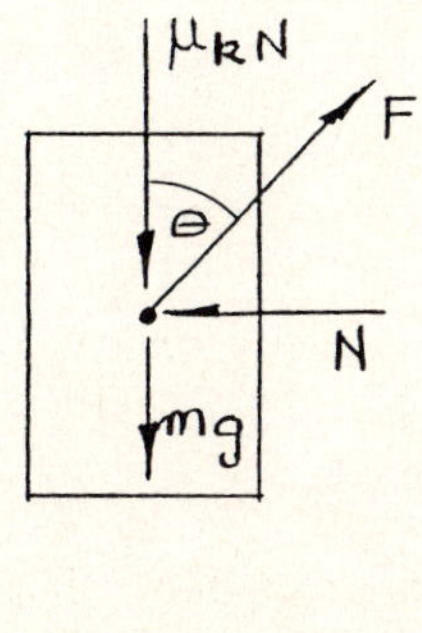

Figure 3.5 Free Body Diagram of the Collar

Resolving Horizontally

Forces to the left = Forces to the right

$F \sin \theta = N$

Frictional Force $= \mu_k N$ (3.7)

Applying Newton's 2nd law in the vertical direction (ie upwards)

$F \cos \theta - \mu_k N - mg = ma$

but $\theta = \mathrm{k}t$

$\therefore F \cos (\mathrm{k}t) - \mu_k F \sin (\mathrm{k}t) - mg = ma$ (3.8)

but $a = \dfrac{\mathrm{d}v}{\mathrm{d}t}$, so that equation (3.8) becomes:

$$\int_o^t [F \cos (\mathrm{k}t) - \mu_k F \sin (\mathrm{k}t) - mg]\, \mathrm{d}t = m \int_o^v \mathrm{d}v$$

$$\text{or } \left[\frac{F}{\mathrm{k}} \sin (\mathrm{k}t) + \frac{\mu_k}{\mathrm{k}} F \cos (\mathrm{k}t) - mgt\right]_o^t = mv$$

$$\text{or } \left[\frac{F}{\mathrm{k}} \sin (\mathrm{k}t) + \frac{\mu_k F}{\mathrm{k}} \cos (\mathrm{k}t) - mgt\right] - \left[\frac{\mu_k F}{\mathrm{k}}\right] = mv$$

$$\text{or } \frac{F}{\mathrm{k}} \{\sin (\mathrm{k}t) + \mu_k [\cos (\mathrm{k}t) - 1]\} = m (v + gt) \tag{3.9}$$

Now when $\mathrm{k}t = \pi/3$, $v = o$, hence equation (3.9) becomes

$$\frac{F}{\mathrm{k}} \left\{ \sin \left(\frac{\pi}{3}\right) + \mu_k \left[\cos \left(\frac{\pi}{3}\right) - 1\right]\right\} = mg \cdot \frac{\pi}{3\mathrm{k}}$$

$$\text{or } F [0.866 + \mu_k (0.5 - 1)] = \pi\, mg/3$$

$$\text{or } F = \pi\, mg / [(0.866 - 0.5\, \mu_k) \times 3]$$

Example 3.4

A system of inter-connected particles is shown in Figure 3.6. Assuming that the mass of the pulleys and cord is negligible and neglecting frictional effects, determine the acceleration of the mass 'B'.

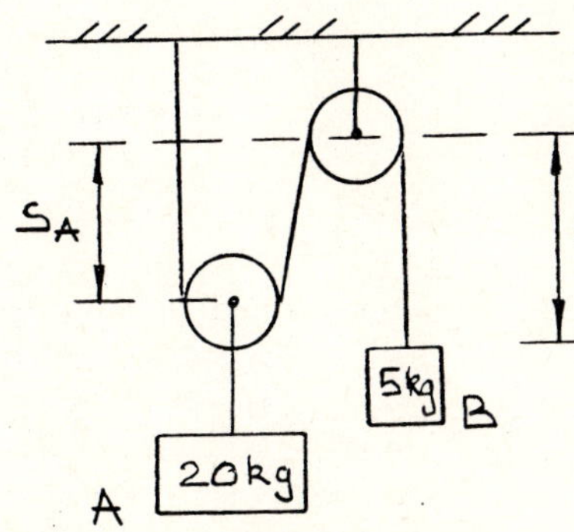

Figure 3.6

The free body diagrams for the masses 'A' and 'B' are shown in Figure 3.7.

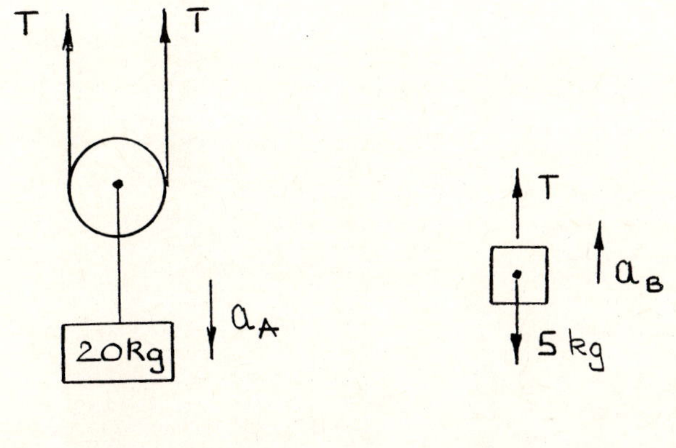

(a) Mass A (a) Mass B

Figure 3.7 Free Body Diagrams for the Masses 'A' and 'B'

From Figure 3.6

$$2s_A + s_B + C = L$$

$$2\dot{s}_A + \dot{s}_B = 0$$

and $2\ddot{s}_A + \ddot{s}_B = 0$

but $a_A = \ddot{s}_A$

and $a_B = -\ddot{s}_B$

$$\therefore a_B = 2a_A \tag{3.10}$$

Applying Newton's 2nd law of motion to mass 'A', in the direction of a_A, we get

$$20\,g - 2T = 20\,a_A$$

or $2T = 20\,(g - a_A)$

or $T = 10\,(g - a_A)$ *(3.11)*

Applying Newton's 2nd law of motion to mass 'B', in the direction of a_B, we get

$$T - 5\,g = 5\,a_B$$

or $T = 5\,(a_B + g)$ (3.12)

Substituting equation (3.10) into equation (3.12), we get

$$T = 5\,(g + 2\,a_A) \tag{3.13}$$

Equating (3.11) and (3.13), we get

$$10\,(g - a_A) = 5\,(g + 2\,a_A)$$

or $10g - 5g = 10\,a_A + 10\,a_A$

or $20\,a_A = 5g$

or $a_A = 0.491$ m/s^2

Hence from equation (3.10)

$$a_B = 0.981 \text{ m/s}^2$$

Problems 3.1

1. A crate of mass '*m*' travels along a rough horizontal surface with an initial speed of 5 m/s. If the co-efficient of kinetic friction, namely μ_k, is 0.3, determine the time and distance for the crate to come to rest from this initial speed.

[1.699s, 4.247 m]

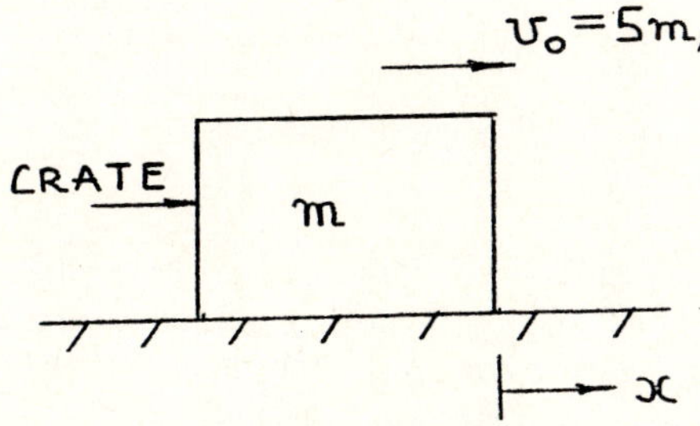

Figure 3.8

2. If the crate of the previous problem were projected down a slope of a similar surface, but with an angle of 10°, determine the time and distance for the crate to come to rest.

[4,184s, 10.46 m]

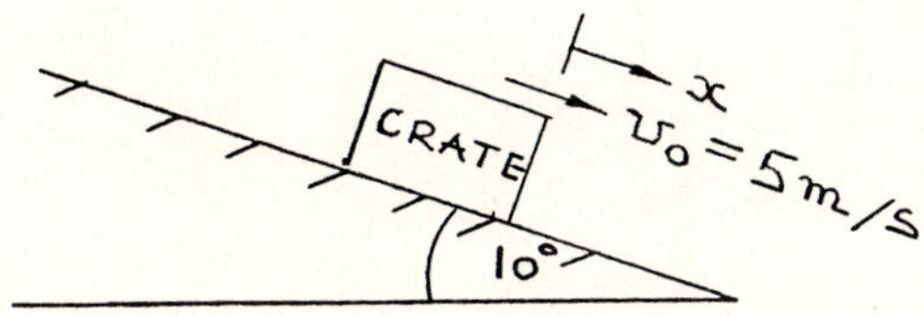

Figure 3.9

3. A lorry travelling on a horizontal surface carries a crate, as shown in Figure 3.10. If the lorry is travelling at 80 km/h, what is the shortest distance that the lorry can stop in, so that the crate will not move forwards. It may be assumed that the coefficient of static friction between the crate and the lorry, namely μ_s, is 0.35.

[71.9 m]

4. If the lorry of the previous problem stopped in 60 m, will the crate strike the back of the lorry, if the coefficient of kinetic friction between the crate and the lorry, namely μ_k, is 0.3? If the crate strikes the back of the lorry, what speed will the crate have just before this happens?

[yes, 6.32 m/s]

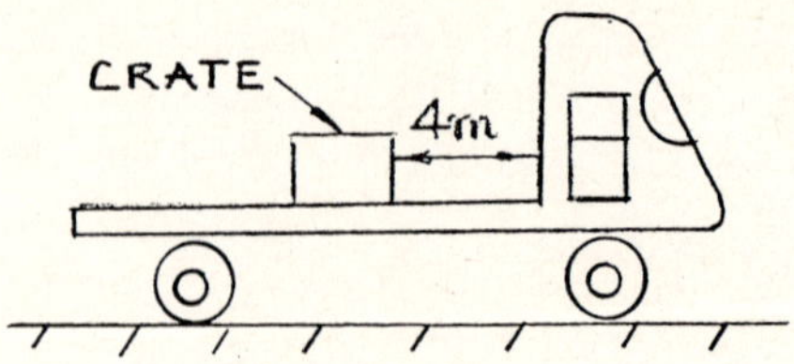

Figure 3.10

5. A man sitting in a cart, pulls himself and the cart up an incline of 10°, with a force of 200 N, by the method shown in Figure 3.11. If the total mass of the man plus cart is 120 kg, determine the acceleration of the man and cart up the slope. Neglect frictional effects and the mass of the pulleys and rope.

[3.3 m/s^2]

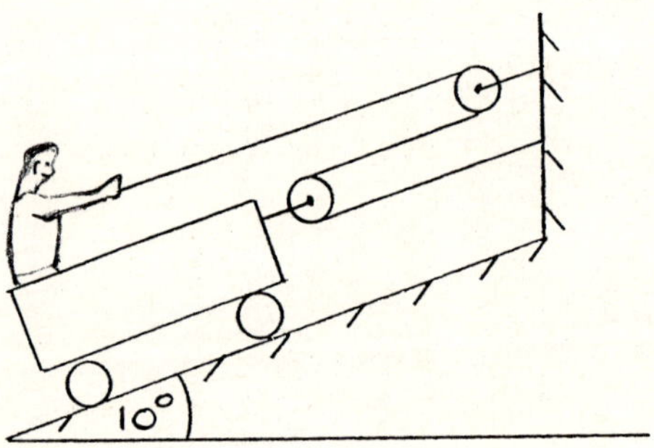

Figure 3.11

6. In the block and tackle system shown in Figure 3.12, the 5 kg mass (B) raises the 10 kg mass (A) from rest. Determine the acceleration of the two masses and the tension in cords 1 and 2.

[1.09 m/s^2, 4.36 m/s^2, 109 N, 27.25 N]

7. A projectile of mass 5 kg is fired vertically into the air with an initial velocity of 60 m/s. If the resistance to motion of the projectile, caused by the atmosphere, namely R, is given by $0.02\ v^2$ N, determine the height that the projectile will rise to, where v = the velocity of the projectile at any time.

[112.9 m]

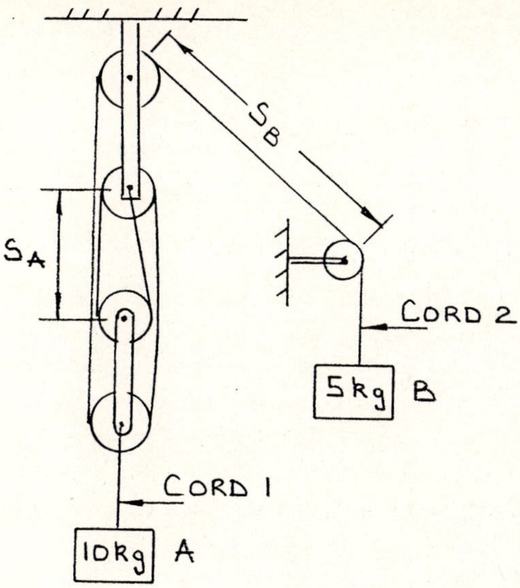

Figure 3.12

8. An aircraft of mass 200 Mg is travelling horizontally at uniform speed of 400 km/h, when its engines provide an additional horizontal thrust T, where $T = 150 + 1.5t^2$ kN. Assuming that when t = o, v = 400 km/h, determine the value of 'v' when t = 10s. Neglect the effects of air resistance.

[436 km/h]

3.4 EQUATIONS OF MOTIONS IN OTHER CO-ORDINATES

In **normal and tangential co-ordinates**, namely the n-t co-ordinate system of Chapter 2, the equations of motion are:

$$\Sigma F_n = m\bar{a}_n \tag{3.14}$$

$$\Sigma F_t = m\bar{a}_t \tag{3.15}$$

where ΣF_n = the resultant force in the 'n' direction

= the centripetal force

ΣF_t = the resultant force in the 't' direction

$\bar{a}_n$ = acceleration in the 'n' direction

$= \rho\,\dot{\theta}^2 = v^2/\rho$

$\bar{a}_t$ = acceleration in the 't' direction

$= \dot{v}$

In **polar co-ordinates**, namely the r-θ co-ordinate system of Chapter 2, the equations of motion are:

$$\Sigma F_r = m\bar{a}_r \qquad (3.16)$$

$$\Sigma F_\theta = m\bar{a}_\theta \qquad (3.17)$$

where ΣF_r = the resultant force in the 'r' direction

ΣF_θ = the resultant force in the 'θ' direction

$\bar{a}_r$ = acceleration in the 'r' direction

$= \ddot{r} - r\dot{\theta}^2$

$\bar{a}_\theta$ = acceleration in the 'θ' direction

$= r\ddot{\theta} + 2\,\dot{r}\,\dot{\theta}$

3.5 D'ALEMBERT'S PRINCIPLE

D'Alembert's principle is used to simplify analysis by replacing the accelerating force by an equal and opposite force. In so doing the dynamic system is in equilibrium, hence, solution of the problem can be achieved by using the principles of statics.

A typical example of this is to replace the **centripetal force** by an equal and opposite imaginary force, namely the **centrifugal force.** In practice the centrifugal force does not exist, but by replacing the centripetal force by the centrifugal force, the dynamic analysis is simplified.

Example 3.5

A racing car is travelling over the brow of the hill, where the radius of curvature ρ = 30 m. Determine the maximum possible speed of the racing car when it is on the brow of the hill, so that it does not leave the road.

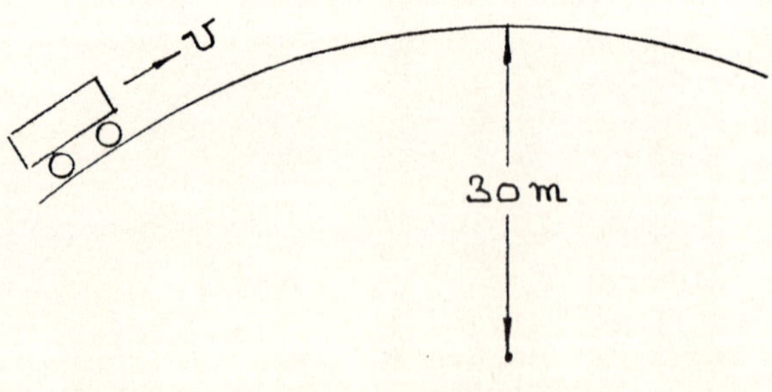

Figure 3.13

$$\text{Now } \overline{a_n} = \frac{v^2}{\rho}$$

the free body diagram of the racing car is shown in Figure 3.14.

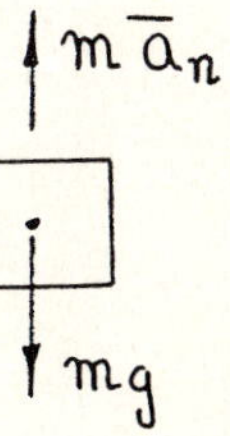

Figure 3.14 Free Body Diagram of Car

$$m\overline{a_n} = mg$$

$$\overline{a_n} = \frac{v^2}{\rho} = g$$

$$\text{or } v = \sqrt{(g\rho)} = \sqrt{(9.81 \times 30)}$$

$$= 17.155 \text{ m/s}$$

$$= 17.155 \times 3.6 \text{ km/h}$$

$$v = 61.76 \text{ km/h}$$

Example 3.6

A motor car is travelling on a level road, which curves in the form of an 'S' bend, as shown in Figure 3.15. When the car is at the point 'A', its speed is 80 km/h and to negotiate the bend, the car's speed is uniformly reduced to 40 km/h at the point 'C'. If the mass of the car is 1000 kg, determine the magnitude of the horizontal tangential force on the car's tyres at the positions 'A', 'B' and 'C'.

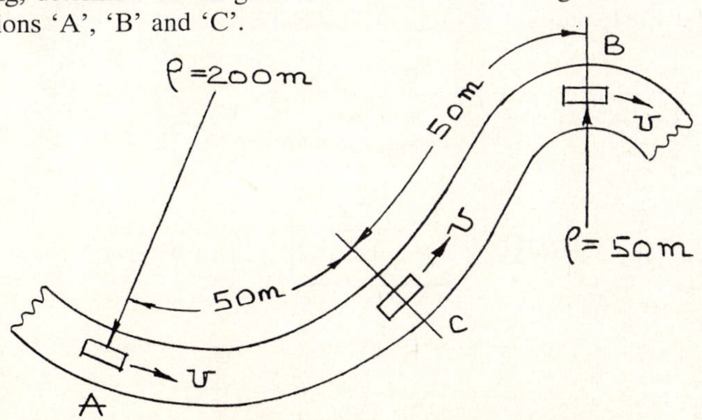

Figure 3.15 'S' Bend in Road

$$v_o = \frac{80}{3.6} = 22.22\,\text{m/s}$$

$$v = \frac{40}{3.6} = 11.11\,\text{m/s}$$

$$v^2 = v_o^2 + 2a_t s$$

$$11.11^2 = 22.22^2 + 2a_t \,.\, 100$$

$$\text{or } 200\, a_t = -370.27$$

$$\therefore\ a_t = -1.851\,\text{m/s}^2 = \text{a constant}$$

From equation (3.15)

$$F_t = m\, a_t = 1000 \text{ x } 1.851 = 1851 \text{ N} = \text{a constant}$$

In the **normal direction**

$$@\ (\text{A}),\ \ a_n = v^2/\rho = 22.22^2/200$$

$$= 2.469 \text{ m/s}^2$$

From equation (3.14)

$$F_n = m\, a_n = 2468.6 \text{ N}$$

$$@\ (\text{B}),\ \ a_n = \text{o} \quad \text{or} \quad F_n = \text{o}$$

$$@\ (\text{C}),\ \ a_n = 11.11^2/50 = 2.469 \text{ m/s}^2$$

$$\text{and } F_n = 2468.6 \text{ N}$$

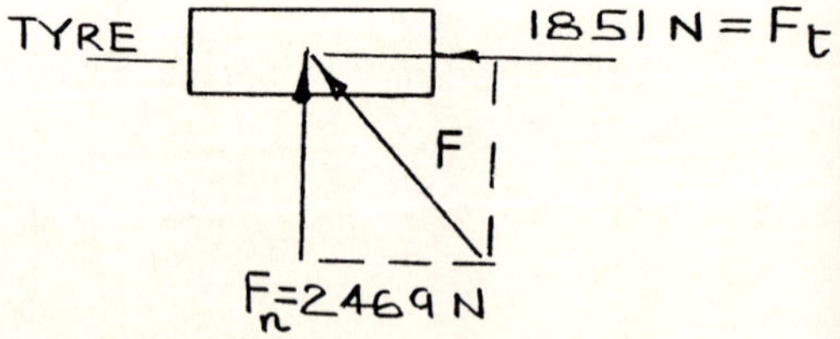

Horizontal Tangential Forces on Tyres

$$@\ (\text{A}),\ \ F = \sqrt{(1851^2 + 2468.6^2)} = 3085.5\text{N}$$

$$@\ (\text{B}),\ \ F = 1851 \text{ N}$$

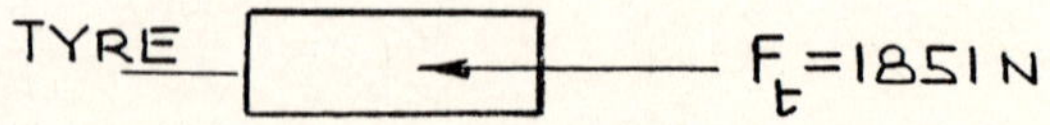

$$@\ (\text{C}),\ \ F = \sqrt{(1851^2 + 2468.6^2)} = 3085.5\text{N}$$

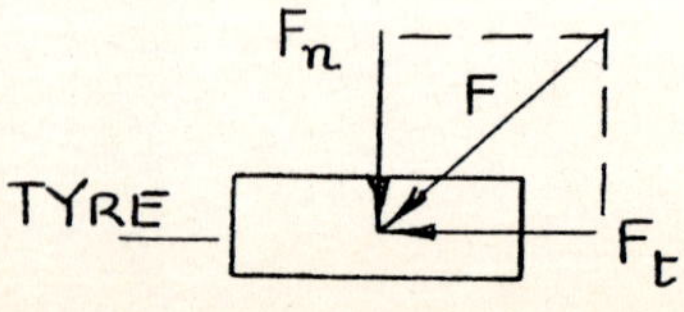

Example 3.7

A four wheeled car of mass 1000 kg is negotiating a bend in a road of radius 150 m at a uniform speed of 100 km/h. If the spacing between the wheels of the car is 1.8 m and the car's centre of gravity is 1.3 m above the surface of the road; determine the minimum required value of the coefficient of static friction, namely μ_s, between the tyres and the road surface that will prevent slipping. Determine the loads that occur on the tyres, assuming that the loads are evenly distributed between the front and rear tyres.

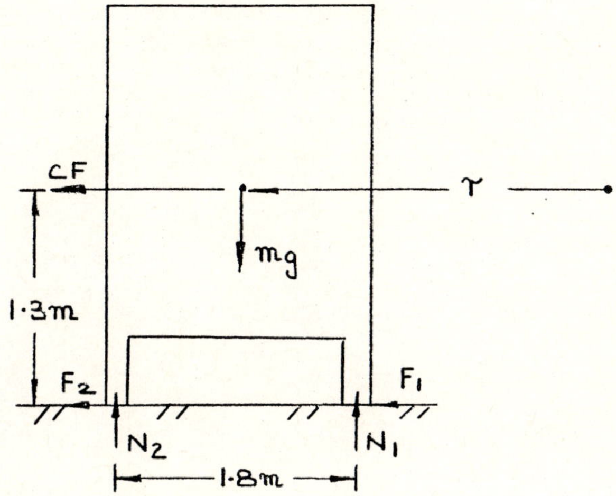

Figure 3.16

$$CF = \text{centrifugal force} = mv^2/r$$

$$v = \frac{100}{3.6} = 27.78\,\text{m/s}$$

$$\therefore \quad CF = \frac{1000 \times 27.78^2}{150} = 5144.9\,\text{N}$$

By assuming a centrifugal force, the problem can be solved by statics.

Resolving Vertically

$$N_1 + N_2 = mg = 1000 \times 9.81$$

$$\text{or } N_1 + N_2 = 9810 \text{ N} \tag{3.17}$$

Resolving Horizontally

$$CF = F_1 + F_2$$

$$\text{or } 5144.9 = \mu_s (N_1 + N_2) \tag{3.18}$$

Substituting equation (3.17) into (3.18)

$$\mu = 5144.9/9810 = 0.525$$

Moments about wheels 1

$$N_2 \text{ x } 1.8 = CF \text{ x } 1.3 + mg \text{ x } 0.9$$

$$1.8\, N_2 = 5144.9 \text{ x } 1.3 + 9810 \text{ x } 0.9$$

$$= 15517.4$$

$$\text{or } N_2 = 8621 \text{ N} \tag{3.19}$$

Substituting equation (3.19) into equation (3.17), we get

$$N_1 = 9810 - 8621 = 1189 \text{ N} \tag{3.20}$$

$$\text{Now } F_1 = \mu_s N_1 \text{ and } F_2 = \mu_s N_2$$

$$\therefore F_1 = 624.4 \text{N}$$

$$\text{and } F_2 = 4526 \text{N}$$

Forces on Tyres

F_{T1} = force on tyre 1

$$= \sqrt{\left(F_1^2 + N_1^2\right)} = 1343 \text{N}$$

F_{T2} = force on tyre 2

$$= \sqrt{\left(F_2^2 + N_2^2\right)} = 9737 \text{N}$$

Example 3.8

A 1 kg mass is subjected to a variable force, so that it moves on a smooth horizontal surface according to the expressions:

$r = 3t^2$ m and $\theta = 0.2t$ rads

Determine the magnitude of the force 'F' on the mass after a period of 2 seconds.

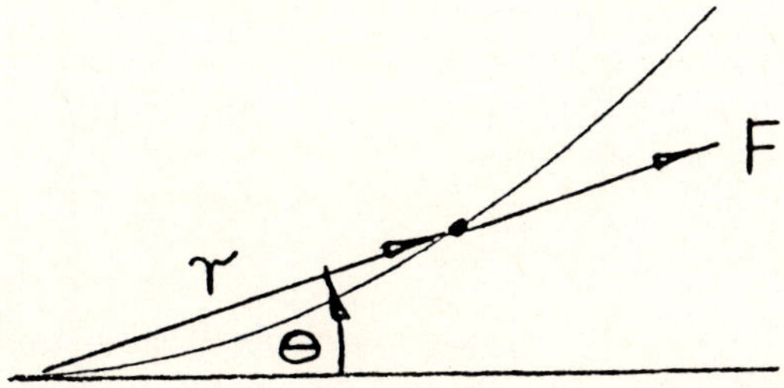

Figure 3.17

Now $r = 3t^2$

$\dot{r} = 6t$

$\ddot{r} = 6 \text{ m/s}^2$

$\theta = 0.2t$

$\dot{\theta} = 0.2 \text{ rads/s}$

$\ddot{\theta} = 0$

Now $a_r = \ddot{r} - r\,\dot{\theta}^2$

$= 6 - 3 \times 2^2 \times 0.2^2$

$a_r = 5.52 \text{ m/s}^2$

$a_\theta = r\ddot{\theta} + 2\,\dot{r}\,\dot{\theta}$

$= 3 \times 2^2 \times 0 + 2 \times 6 \times 2 \times 0.2$

$a_\theta = 4.8 \text{ m/s}^2$

$F_r = m\,a_r = 1 \times 5.52 = 5.52 \text{ N}$

$F_\theta = m\,a_\theta = 1 \times 4.8 = 4.8 \text{ N}$

$F = \sqrt{(5.52^2 + 4.8^2)} = 7.315 \text{ N}$

Problems 3.2

1. A four wheeled car of mass 1000 kg is negotiating a bend in a road of radius 100 m, at a uniform speed of 80 km/h. If the spacing between the car wheels is 2 m and the height of the centre of gravity of the car above the road surface is 1.5 m, determine the forces on the car tyres, assuming that the coefficient of static friction between the car tyres and the surface of the road, namely μ_s is sufficient to prevent slipping and also that the loads on the front tyres are the same as those on the back tyres.

   [672.8 N, 4618N]

2. If the car of problem 1 were negotiating a similar bend in a road, but that the bend is banked, as shown in Figure 3.18, determine the forces in the tyres.

   [4521 N, 6461 N]

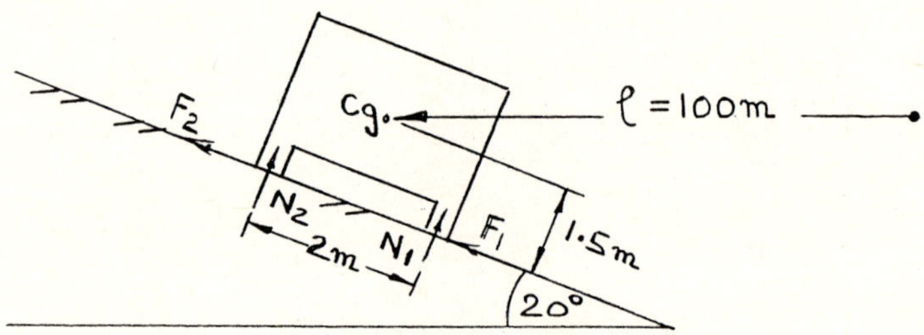

Figure 3.18

3. A skier of mass 80 kg is travelling at a uniform speed of 40 km/h over the curved surface shown in Figure 3.19. If the skier is treated as a particle, determine the force exerted by the ground on the skier's feet at the base of the curved surface, where the radius of curvature is 10 m.

   [1772 N]

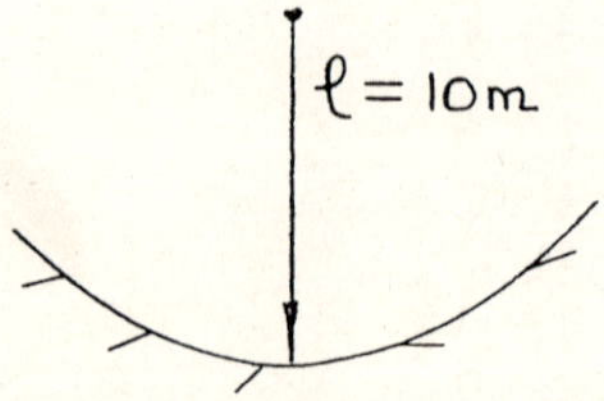

Figure 3.19

4. If the skier of the previous problem skied over a hump with a radius of curvature of 20 m at 40 km/h, what would be the force exerted by the surface on the skier's feet. At what speed will the skier leave the surface?

[291.1 N, 50.4 km/h]

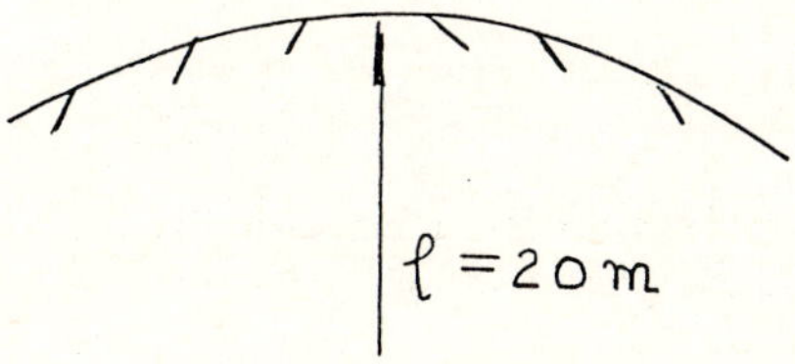

Figure 3.20

5. A conical pendulum rotates with a constant angular velocity of $\dot{\theta} = \omega$. Determine the height '*h*' and the tension '*T*' in the cord. Neglect the mass of the cord.

[g/ω^2, $m\omega^2 \ell$]

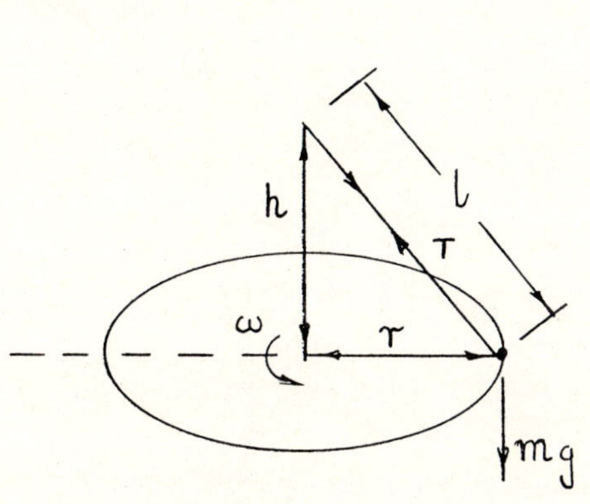

Figure 3.21 Conical Pendulum

6. A rod of uniform circular section rotates abut 'O' in a counter-clockwise direction, in the manner shown in Figure 3.22, with an angular velocity of $\dot{\theta}$ = 4 rads/s. Sliding radially on the rod is an annular sleeve, which is travelling inwards with a relative velocity to the shaft of 1.5 m/s. When the angle θ = 20°, determine the normal force exerted by the rod on the sleeve, assuming the mass of the sleeve is 0.2 kg.

[-0.557 N]

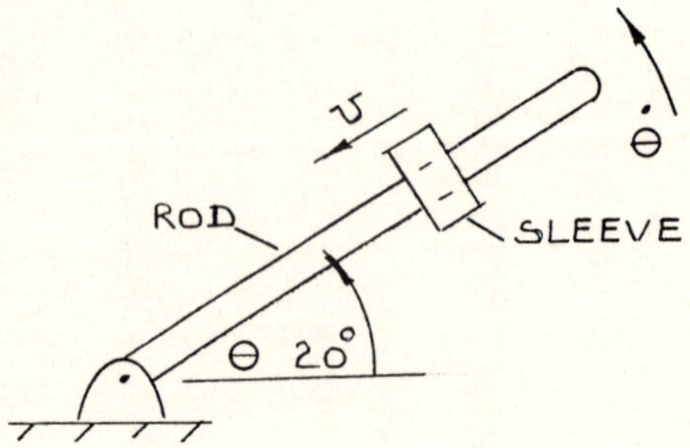

Figure 3.22

7. A plane which is travelling at a constant speed of 200 km/h executes the loop shown in Figure 3.23, where $r = -200 \cos 2\theta$ m, where 'θ' is in radians. If the mass of the pilot is 80 kg, determine the force exerted by the plane's seat on the pilot, when the plane is in position 'A'.

[5398 N]

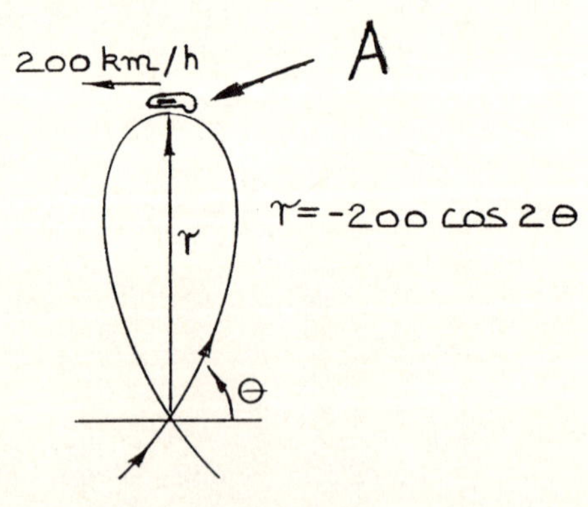

Figure 3.23

8. An artefact consists of a horizontal circular ring of constant circular cross-section. If a small annular collar of mass 'm', which fits snugly on to this ring, as shown in Figure 3.24, is given an initial velocity 'v_o', over what distance will the collar travel before it comes to rest? It may be assumed that the coefficient of kinematic friction between the collar and ring is μ_k.

$$\left[\frac{\rho}{2\mu_k} \quad \ln \left[\frac{v_o^2 + \left(v_o^4 + g^2\ \rho^2\right)^{\frac{1}{2}}}{g\rho}\right]\right]$$

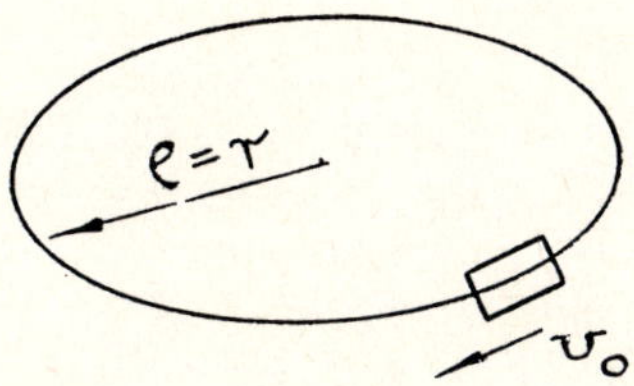

Figure 3.24

CHAPTER FOUR

Kinetics of Particles: Work, Energy and Power

4.1 INTRODUCTION

In this chapter, the terms strain energy and work are defined, together with power and efficiency.

In general, problems in this domain usually involve velocity, force and displacement. A number of worked examples are given, which demonstrate how the principles of work done and mechanical energy can be used for solving problems involving the dynamics of particles.

The conservation of energy is also defined.

4.2 WORK DONE

The work done by a force 'F' moving over a distance 'x', where both 'F' and 'x' are in the same direction is given by:

$$U = F \cdot x \tag{4.1}$$

where U = work done

The units of work done are Joules (J), where

$$1\,\mathrm{J} = 1\,\mathrm{N} \cdot \mathrm{m} \tag{4.2}$$

Although work done is a scalar quantity, the general expression for work done is given by the dot product of two vector quantities, as follows:

$$\mathrm{d}U = \text{work done} = \overline{F} \cdot \mathrm{d}\overline{s} \tag{4.3}$$

where $\overline{F}$ = a vector force
$\mathrm{d}\overline{s}$ = a vector displacement (see Figure 4.1).

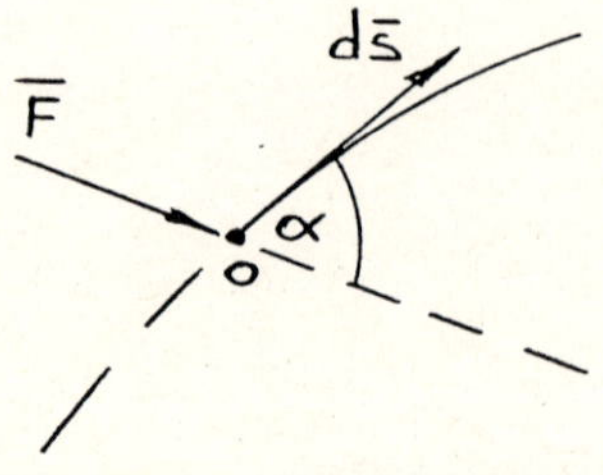

Figure 4.1

From Figure 4.1, it can be seen that the angle between the direction of $\bar{F}$ and $d\bar{s}$ is α, so that

$$dU = F\cos\alpha \,.\, d\bar{s} \tag{4.4}$$

where

$F \cos\alpha = F_t =$ tangential component of $\bar{F}$

It should be noted that the normal component of $\bar{F}$, namely F_n is equal to $F \sin\alpha$, and it does not do work.

For the directions of F_n and F_t, see Figure 4.2.

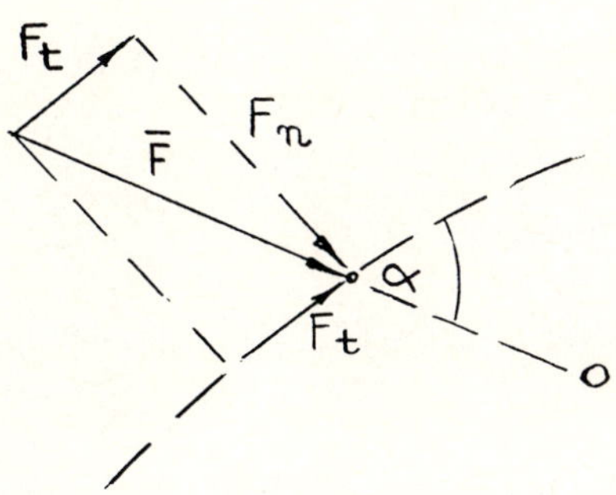

Figure 4.2 Normal and Tangential Components of $\bar{F}$

Alternatively, from Figure 4.2,

$$dU = \bar{F} \,.\, ds \,.\, \cos\alpha \tag{4.5}$$

which can be expressed as: "that the work done is equal to the vector force multiplied by the component of displacement in the direction of vector force"; this expression can be seen to be identical to equation (4.4).

NB It should be noted that when $0 \le \alpha < 90°$, the work is said to be **positive**.

Similarly, when $90° < \alpha \le 180°$, the work done is said to be **negative.**

That is, if the displacement is in the opposite direction of the force, the work done is negative and vice versa for positive work done.

When $\alpha = 90°$, $\cos 90° = 0$ and there is no work done.

In general,

$$U = \int \overline{F} \cdot d\overline{s} \tag{4.6}$$

In **rectangular co-ordinates**, equation (4.6) becomes:

$$U = \int (F_x \cdot dx + F_y \cdot dy + F_z \cdot dz) \tag{4.7}$$

In **normal and tangential co-ordinates**, equation (4.6) becomes:

$$U = \int F_t \cdot ds \tag{4.8}$$

Equations (4.7) and (4.8) can be solved either explicitly or graphically or numerically, depending on what is known about the motion.

4.3 ELASTIC STRAIN ENERGY

The units for energy including strain energy are the same as that for work done, because the equation for strain energy can be obtained from equation (4.6) as follows.

Consider the load-deflection relationship for a linear elastic spring, as shown in Figure 4.3.

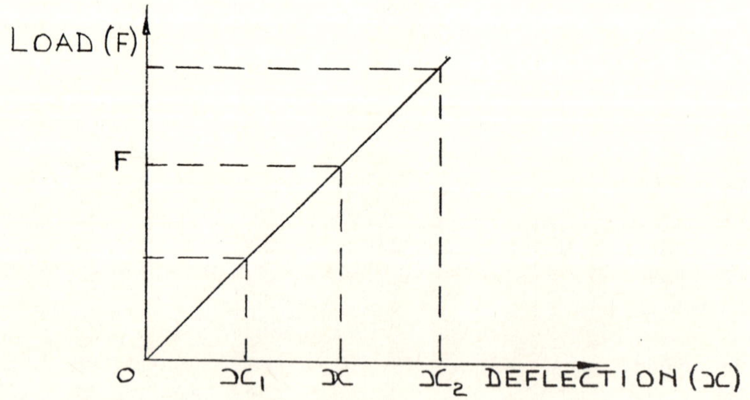

Figure 4.3 Load-Deflection Relationship for a Spring

Robert Hooke (1635-1703) found that for many structural materials, that the applied load on the material was proportional to the resulting deflection. That is, if twice the load were applied to a structure, the structure would deflect twice as much, so that, in general,

$$F = \mathrm{k}x \tag{4.9}$$

where

k = stiffness = a constant dependent on material and geometrical properties

x = deflection due to the force 'F'

Now the work done in compressing or extending the spring is positive, because both 'F' and 'x' are the same direction and it is given by:

$$U = \int F \,.\, \mathrm{d}x \tag{4.10}$$

Substituting equation (4.9) into equation (4.10)

$$U = \int_{x_1}^{x_2} \mathrm{k}\, x \,.\, \mathrm{d}x$$

$$= \left[\mathrm{k}\,\frac{x^2}{2}\right]_{x_1}^{x_2}$$

$$\text{or } U = \frac{1}{2}\,\mathrm{k}\left(x_2^2 - x_1^2\right)$$

= strain energy stored (4.11)

where

x_2 = final deflection of spring

x_1 = intial deflection of spring

If the spring is loaded from an initial zero deflection, then

$$U = \frac{1}{2}\,\mathrm{k}x^2 \tag{4.12}$$

= strain energy stored

where

x = deflection of the spring due to 'F'

Strain energy can be described as the mechanical energy in a structure due to external loadings.

4.4 KINETIC ENERGY (*KE*)

Kinetic energy can be described as energy due to dynamical motion; its units are in Joules.

Now, in general terms,

$$\text{Work done} = U = \int \bar{F} \,.\, \mathrm{d}\bar{r}, \tag{4.13}$$

where

$\bar{r}$ = a positional vector

$$\text{but } \bar{F} = m\bar{a} \tag{4.14}$$

Substituting equation (4.14) into equation (4.13)

$$U = \int m\bar{a} \,.\, \mathrm{d}\bar{r}$$

In terms of tangential co-ordinates

$$\bar{a} = a_t \text{ and } \mathrm{d}\bar{r} = \mathrm{d}s$$

$$\therefore U = \int m \,.\, a_t \,.\, \mathrm{d}s \tag{4.15}$$

$$\text{but } a_t = v \,.\, \frac{\mathrm{d}v}{\mathrm{d}s}$$

$$\therefore U = \int_{v_1}^{v_2} m \,.\, v \,.\, \mathrm{d}v = \left[\frac{mv^2}{2}\right]_{v_1}^{v_2} \tag{4.16}$$

$$U = \frac{m}{2}\left(v_2^2 - v_1^2\right) = \text{kinetic energy} = KE \tag{4.17}$$

where

v_1 = intial velocity

v_2 = final velocity

If $v_1 = 0$

$$KE = \frac{1}{2} m v^2 \tag{4.18}$$

where

v = velocity at any time

4.5 POWER (P)

Power is defined as the rate of doing work. It is a scalar quantity and its units are in watts (W).

$$\text{Power } P = \overline{F} \cdot \overline{v} \tag{4.19}$$

NB 1 W = 1 N m/s = 1 J/s

4.6 MECHANICAL EFFICIENCY (η)

Mechanical efficiency =

$$\eta = \frac{\text{Power output}}{\text{Power input}} \text{ x } 100\% \tag{4.20}$$

$$= \frac{P_{output}}{P_{input}} \; x \; 100\%$$

4.7 POTENTIAL ENERGY (PE)

Potential energy (*PE*) can be regarded as stored mechanical energy. Its units are in Joules. One form of potential energy is elastic strain energy as described in Section 4.3 and another form of PE is gravitational *PE*. In this case, negative work is done in raising a mass '*m*' up to a height '*h*' above the ground, so that the potential energy above the ground is given by:

$$PE = mgh \tag{4.21}$$

NB Gravitational *PE* is always relative to a certain position, which may not be the ground.

4.8 CONSERVATION OF ENERGY

The law of the conservation of energy is that ‘Energy can neither be created nor destroyed’. It should be emphasised that although energy cannot be created or destroyed, it can change its form, including into heat, light and sound energy. These latter forms of energy are difficult to consider and are beyond the scope of the present book, thus, they will usually not be considered in the calculations of the present book.

Energy can be released from many different processes, including through the burning of fossil fuels and also through chemical and nuclear reactions. At present, the most common method of changing energy stored in chemical form to mechanical energy is through the burning of fossil fuels such as gas, oil and coal. In the case of oil, it is very often refined to other forms of stored energy, such as petrol and diesel fuel prior to burning them in an internal combustion engine so that their chemical energy can be converted to give mechanical energy. In some cases, as it is in the case of electrical energy, the energy stored in fossil fuels is converted into electrical energy through electrical generators, which are driven by steam turbines, the steam being produced by burning fossil fuels. As the process of converting the chemical energy stored in fossil fuels into electrical energy is expensive and inefficient, it is best to avoid using electrical energy as a space heater, if at all possible.

NB 1 amp . volt = 1 watt = 1 J/s

= 1 N . m/s

Also 1 calorie is the heat required to raise 1 gram of water by 1°C.

Example 4.1

A suitcase of mass 20 kg is pushed down a chute, with a initial velocity of 5m/s at the point ‘A’. If the co-efficient of kinetic friction between the suitcase and the chute’s surface is 0.35, and the angle that the chute makes with the horizontal is 20°, determine the velocity of the suitcase at the point ‘B’.

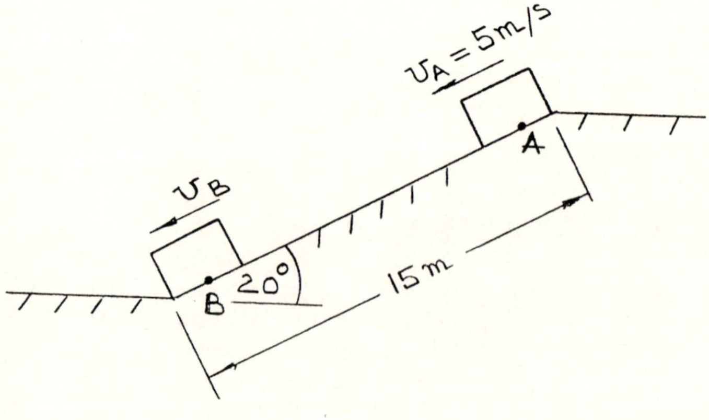

Figure 4.4

The free body diagram is shown in Figure 4.5.

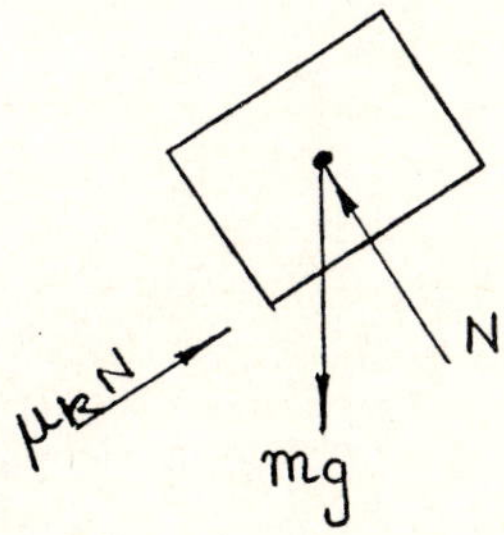

Figure 4.5

Resolving perpendicular to the surface, we get

$$N = mg \cos 20°$$

$$= 20 \text{ x } 9.81 \text{ x } 0.9397$$

$$= 184.4 \text{ N}$$

Hence the frictional force parallel to the surface,

$$F_r = \mu_k N = 0.35 \text{ x } 184.4$$

$$F_r = 64.5 \text{ N}$$

$$\text{Loss of } PE = mg \text{ x } 15 \sin \alpha$$

$$= 20 \text{ x } 9.81 \text{ x } 15 \text{ x } \sin 20°$$

$$= 1006.6 \text{ J}$$

$$\text{`Loss' of } KE = \frac{m}{2}\left(v_A^2 - v_B^2\right)$$

$$= \frac{20}{2}\left(5^2 - v_B^2\right)$$

$$= 10\left(25 - v_B^2\right)$$

Energy absorbed by the frictional force

$$= \mu_k \, . \, N \times 15$$

$$= 64.5 \times 15$$

$$= 967.5$$

From the conservation of energy

Loss of *PE* + 'Loss' of *KE* = Energy absorbed in friction

$$\text{or } 1006.6 + 2\left(25 - v_B^2\right) = 967.5$$

$$\text{or } 25 - v_B^2 = \frac{-39.1}{2}$$

$$v_B^2 = 44.55$$

$$v_B = 6.675 \text{ m/s}$$

ie there is a gain in *KE*

Example 4.2

If the suitcase of Example 4.1 were assumed to move on to the horizontal part of the chute, immediately after reaching the point 'B', how far will the suitcase travel before coming to rest?

The free body diagram is shown in Figure 4.6.

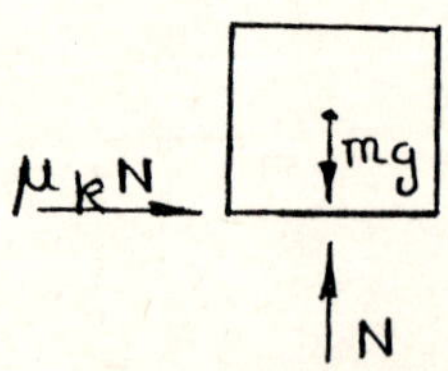

Figure 4.6

Resolving vertically, we get

$$N = mg = 20 \times 9.81 = 196.2 \text{ N}$$

Hence, frictional force

$$= F_r = \mu_K N = 68.67 \text{ N}$$

$$\text{loss of } KE = \frac{m}{2}\left(v_B^2 - 0\right)$$

$$= \frac{20}{2} x\ 6.675^2$$

$$\text{loss of } KE = 445.6 \text{ J}$$

Frictional energy absorbed

$$= 68.67S$$

From the conservation of energy

$$\text{loss of } KE = \text{Frictional energy absorbed}$$

$$445.6 = 68.67\,S$$

$$\therefore S = \text{distance travelled before coming to rest}$$

$$= 6.49 \text{ m}$$

NB It should be noted that Examples 4.1 and 4.2 could have been solved from considerations of force, mass and acceleration.

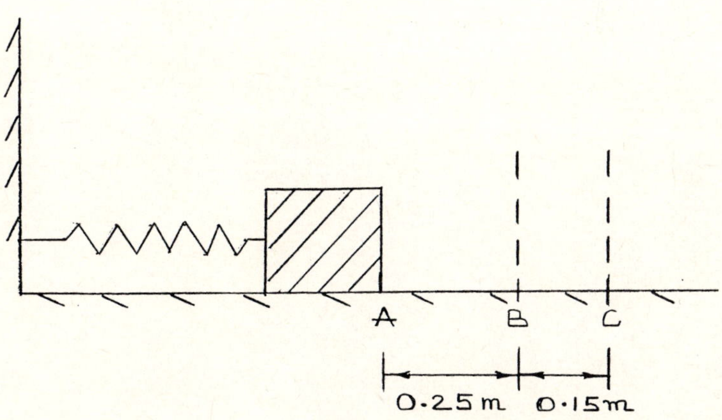

Figure 4.7

Example 4.3

A block of mass 5 kg lies on a rough horizontal surface alongside a horizontal spring of stiffness 400 N/m at 'B'. The spring is not attached to the block, but the block is then pushed to the left, so that it compresses the spring by 0.25 m at the position 'A'. If the block is suddenly released what will be its velocity at 'C', assuming that the coefficient of kinetic friction, namely μ_k is 0.3.

$$\text{Loss of strain energy} = \frac{1}{2}\,k\,x^2 = \frac{1}{2}\,.\,400\,.\,0.25^2$$

$$= 12.5 \text{ J}$$

$$\text{Gain in } KE = \frac{mv_c^2}{2} = \frac{5\,v_c^2}{2} = 2.5\,V_c^2$$

where v_c = velocity at C

Energy absorbed in friction

$$= 0.3 \text{ x } mg \text{ x } (0.25 + 0.15)$$

$$= 5.886 \text{ J}$$

From the conservation of energy

Loss in strain energy = Gain in *KE* + Energy absorbed in friction

$$\text{or } 12.5 = 2.5\,v_c^2 + 5.886$$

$$\text{or } v_c^2 = 2.646$$

$$v_c^2 = 1.627 \text{ m/s}$$

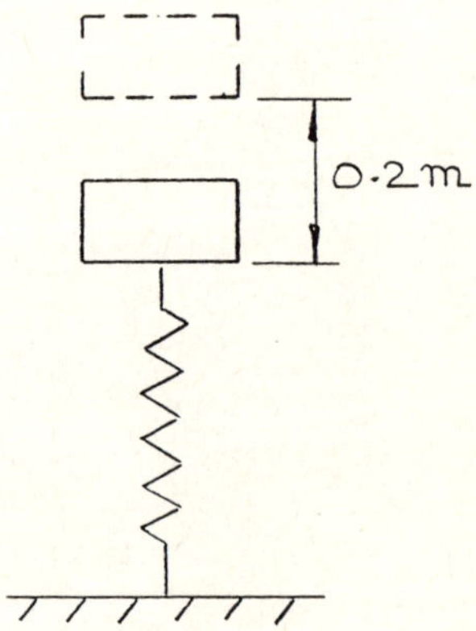

Figure 4.8

Example 4.4

A piece of metal of mass 1.2 kg is placed on top of a vertical spring and causes the spring to deflect by 0.05 m, due to the self mass of the metal. A man than pushes the spring so that it compresses by a further 0.15 m. If the mass is then suddenly released, from rest to what height will the mass rise from its position when the spring is compressed.

The stiffness of the spring,

$$k = \frac{mg}{\delta} = \frac{1.2 \text{ x } 9.81}{0.05} = 78.48 \text{ N/m}$$

where δ = deflection of the spring due to the self weight of the mass.

As the *KE* = 0 at the start and the end, the effects of *KE* can be ignored.

Loss of strain energy = Gain in *PE*

$$\frac{1}{2} \text{ x } 78.48 \text{ x } (0.05 + 0.15)^2 = 1.2 \text{ x } 9.81 \text{ x } h$$

$$\text{or } 1.5696 = 1.2 \text{ x } 9.81 \text{ x } h$$

or h = the height the mass rises

$$h = 0.133 \text{ m}$$

Example 4.5

A block of mass 1.5 kg slides down the smooth curved cylindrical surface shown in Figure 4.9. If the block is given an initial velocity of 0.8 m/s, at what value of 'θ' will the block leave the surface of the cylinder?

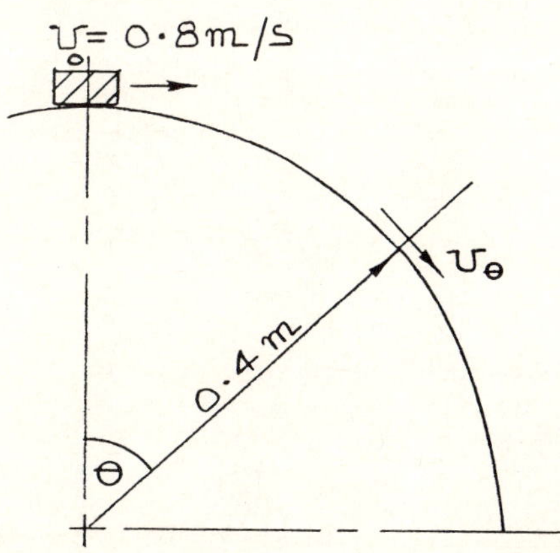

Figure 4.9

To determine v_θ

Loss of *PE* $= mg \times 0.4\,(1 - \cos\theta)$

Gain in *KE* $= m/2\,(v_\theta^2 - v_0^2)$

From the conservation of energy

Loss of *PE* = Gain in *KE*

$$mg \times 0.4\left(1 - \cos\theta\right) = \frac{m}{2}\left(v_\theta^2 - 0.8^2\right)$$

$$\text{or } 7.848\left(1 - \cos\theta\right) = v_\theta^2 - 0.8^2 \qquad (4.22)$$

$$\text{or } v_\theta^2 = 7.848\left(1 - \cos\theta\right) + 0.64$$

The free body diagram is shown in Figure 4.10.

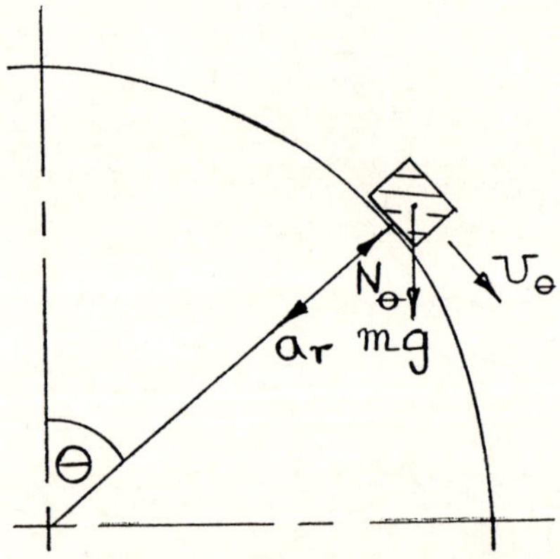

Figure 4.10

Applying Newton's 2nd law in the direction of N_θ

$$N_\theta - mg\cos\theta = -m\,a_n$$

$$= -m\left(\frac{v_\theta^2}{0.4}\right)$$

but when the block leaves the surface, $N_\theta = 0$

$$\therefore\ g\cos\theta = \frac{v_\theta^2}{0.4}$$

$$\text{or } v_\theta^2 = 0.4g\cos\theta \tag{4.23}$$

$$\text{or } v_\theta^2 = 3.924\cos\theta$$

Equating (4.22) and (4.23)

$$8.488 - 7.848\cos\theta = 3.924\cos\theta$$

$$\text{or } \cos\theta = 8.488 / 11.772$$

$$\theta = 43.86°$$

Example 4.6

A system of interconnected particles is shown in Figure 4.11, where block 'A' has a mass of 50 kg and block 'B' has a mass of 100 kg. If the system starts from rest, determine the change in the value for s_B, when the velocity of 'B' reaches 1.5 m/s.

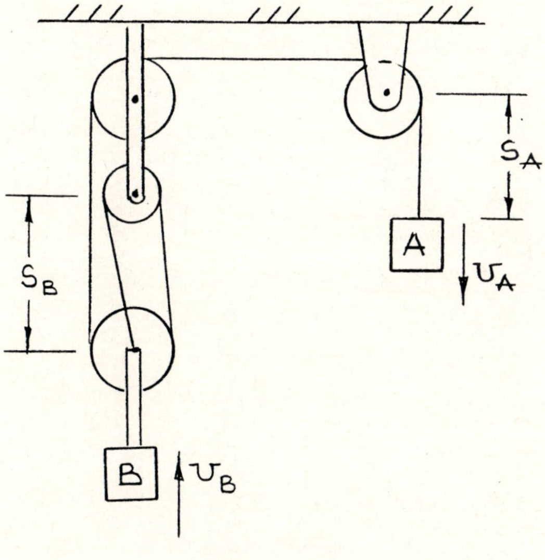

Figure 4.11

$$L = S_A + 3S_B + C$$

$$\frac{dL}{dt} = 0 = \dot{S}_A + 3\dot{S}_B$$

$$\text{or } \dot{S}_A = -3\dot{S}_B$$

$$\text{but } v_A = \dot{S}_A$$

$$\text{and } v_B = -\dot{S}_B$$

$$\therefore v_A = 3v_B = 3 \text{ x } 1.5$$

$$\text{or } v_A = 4.5 \text{ m/s}$$

Loss of *PE* of A $= 50 \text{ x } 9.81 \text{ x } \Delta S_A$

Gain of *PE* of B $= 100 \text{ x } 9.81 \text{ x } \Delta S_B$

Gain of *KE* of A & B $= \dfrac{50 \text{ x } v_A^2}{2} + \dfrac{100}{2} \text{ x } v_B^2$

From the conservation of energy,

Loss of *PE* of 'A' = Gain of *PE* of 'B' + Gain in *KE* of 'A' and 'B'

$$\text{or } 50 \text{ x } 9.81 \text{ x } \Delta S_A - 100 \text{ x } 9.81 \ x \ \Delta S_B$$

$$= \frac{50}{2} \text{ x } v_A^2 + \frac{100}{2} \text{ x } v_B^2$$

$$\text{but } v_B = 1.5 \ \& \ v_A = 4.5$$

Additionally, $\Delta S_A = -3\Delta S_B$

$$\therefore -(1471.5 + 981)\ \Delta S_B = 506.25 + 112.5$$

$$\text{or } \Delta S_B = \frac{-618.75}{2452.5}$$

$$\text{or } \Delta S_B = -0.25 \text{ m}$$

ie 'S_B' is shortening

Problems 4.1

1. A 1 kN force is used to drag a log of mass 200 kg up a rough incline, at a constant velocity. The distance travelled by the log, along the slope is 30 m and the vertical height that the log attains above the bottom of the slope is 10 m. If the log is then released, determine the velocity it attains at the bottom of the slope.

   [9.726 m/s]

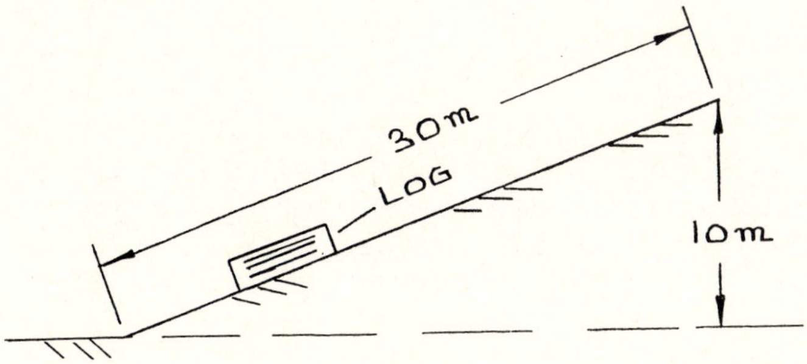

Figure 4.12

2. A truck carries a crate of mass 70 kg. If the truck starts from rest and accelerates uniformly to a speed of 70 km/h in a distance of 80 m, determine the work done by the frictional force between the crate and the truck when,

   (a) μ_s = 0.3 and μ_k = 0.27

   (b) μ_s = 0.23 and μ_k = 0.20

   [1323.2 J, 9123 J]

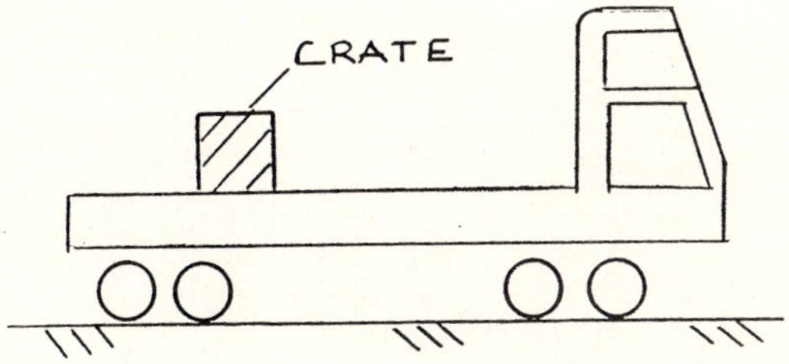

Figure 4.13

3. A man is pedalling a cycle up a gradient of 7%. If the man and the cycle have a combined mass of 80 kg and this combination is travelling at a constant speed of 30 km/h, determine the power output of the man.

   [457.8 W]

4. A man of mass 70 kg runs up a flight of stairs in 3½ seconds, at a constant speed. If the total height of the flight of stairs is 5 m, what power does he generate?

[981 W]

5. An escalator in a departmental store rises by 8 m between two adjacent floors. If the escalator is driven by a motor of 2.5 kW and it can transport 25 people per minute, where the average mass of a person is 60 kg, determine the efficiency of the escalator.

[78.48%]

6. A bullet travelling at a speed of 300 m/s, strikes a wooden door and emerges at a speed of 280 m/s. Determine how many doors will have to be placed flat against each other to completely stop the bullet from emerging.

[6 doors]

7. A hammer of mass 'm' strikes a nail of negligible mass with a velocity 'v'. Neglecting energy losses, and assuming that the hammer and the nail remain together, while the nail travels a distance 'S' into a piece of wood, determine the **average** resistance of the wood to penetration.

[$m\,(v^2 + 2\,gS)\,/\,(2S)$]

8. A ball of mass 0.4 kg is fired up a vertical circular track, as shown in Figure 4.15. If the ball leaves the track when $\theta = 120°$, determine the distance that the spring plunger has to be pulled back to achieve this. It may be assumed that the stiffness of the spring is 600 N/m.

[0.151 m]

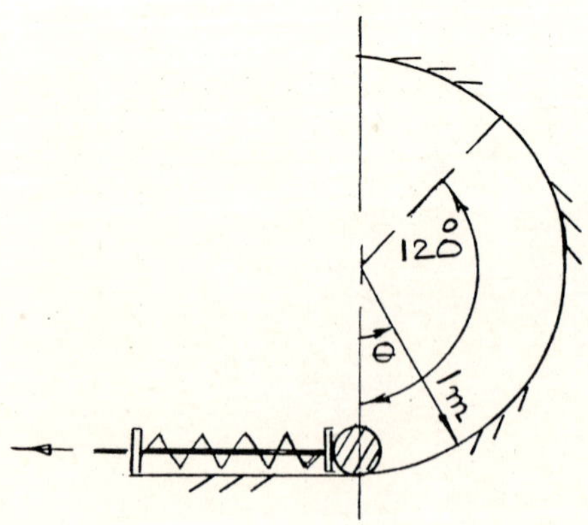

Figure 4.14

9. A boat of mass 3000 kg collides with an elastic bumper which is situated at the dock side. If the resistance of the bumper, namely 'F_R' is given by $F_R = k\,S^3$, determine the required value of 'k' that will cause the bumper to deflect by 0.2 m, when the boat's speed is 0.3 m/s immediately before collision, where S = deflection of bumper.

[337500 N/m^3]

10. A block of mass 2 kg strikes a non-linear elastic spring with an initial velocity of 3 m/s. The spring is called non-linear because its force 'F_s': displacement 'S' relationship is given by $F_s = 800\ S^2$. Determine the velocity of the mass when $S = 0.25$ m.

[2.198 m/s]

11. A girl of mass 35 kg is swinging on the swing shown in Figure 4.15. If the girl's centre of gravity is 1.8 m from the top of the swing and she commences her motion when $\theta = 65°$, determine the force in each of the four posts of the apparatus when $\theta = 0°$. It may be assumed that the posts are perpendicular to the ground in a plane at 90° to that shown.

[204.1 N]

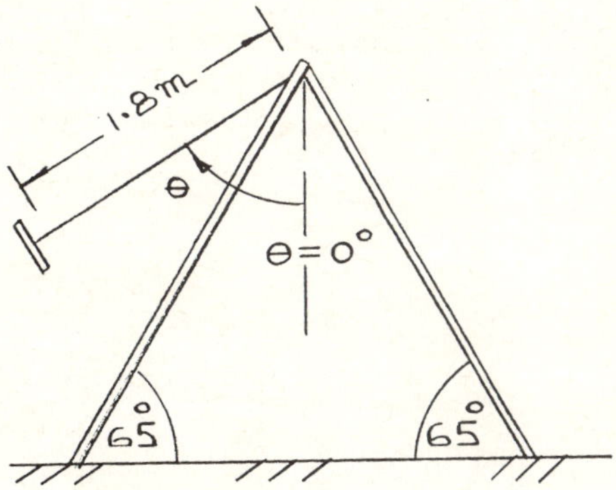

Figure 4.15

CHAPTER FIVE

Kinetics of a Particle: Impulse and Momentum

5.1 INTRODUCTION

In this chapter, definitions will be made of linear and angular impulse and momentum together with the appropriate derivations of their equations. The equations for the conservation of linear and angular momentum will also be defined.

Worked examples will be given on problems involving impulse and momentum and also in problems involving impact.

5.2 LINEAR IMPULSE AND MOMENTUM

Linear momentum can be defined as the product of mass times velocity. It is a vector quantity and therefore its direction is important. Its units are kg.m/s.

Linear impulse can be described as the product of force times time. It too is a vector quantity and its units are in N.s. Now as one Newton equals 1 kg.m/s^2, the units of linear impulse are the same as the units of linear momentum.

The equation for linear momentum can be obtained from the mass: force: acceleration relationship. If the mass is constant,

$$\bar{F} = m\bar{a}$$

$$= m\frac{d\bar{v}}{dt} \qquad (5.1)$$

$$\text{or} \int \bar{F}.\ dt = m \int d\bar{v}$$

so that if a force '$\bar{F}$' is applied from a time 't_1', to a time 't_2', it will cause a mass of value 'm' to increase its velocity from v_1 to v_2, so that equation (5.1) becomes:

$$\int_{t_1}^{t_2} \bar{F}\ dt = m \int_{\bar{v}_1}^{\bar{v}_2} d\bar{v}, \qquad (5.2)$$

or

$$\int_{t_1}^{t_2} \bar{F}\ .\ dt = m\left(\bar{v}_2 - \bar{v}_1\right), \qquad (5.3)$$

where

$$\int_{t_1}^{t_2} \overline{F} \,.\, dt \;=\; \text{linear impulse} \;=\; \overline{I} \tag{5.4}$$

$$\text{and } m\left(\overline{v_2} - \overline{v_1}\right) \;=\; \text{change in linear momentum} \tag{5.5}$$

5.3 PRINCIPLE OF LINEAR MOMENTUM AND IMPULSE

Equation (5.3) can be written:

$$m\,\overline{v_1} + \int_{t_1}^{t_2} \overline{F} \,.\, dt \;=\; m\,\overline{v_2} \tag{5.6}$$

ie the initial momentum at the time 't_1' plus the impulse between 't_1' and 't_2' is equal to the final momentum at time 't_2'. Equation (5.6) is illustrated through the use of Figure 5.1.

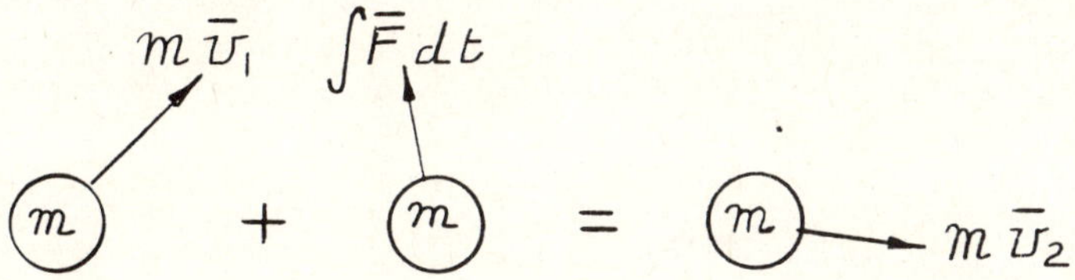

Figure 5.1

From Figure 5.1, it can be seen that the impulse and momentum must be resolved in the appropriate directions for equation (5.6) to apply.

In **three dimensional rectangular co-ordinates**, equation (5.6) becomes:

$$m\,(v_x)_1 + \int_{t_1}^{t_2} F_x \,.\, dt \;=\; m\,(v_x)_2$$

$$m\,(v_y)_1 + \int_{t_1}^{t_2} F_y \,.\, dt \;=\; m\,(v_y)_2 \tag{5.7}$$

$$m\,(v_z)_1 + \int_{t_1}^{t_2} F_z \,.\, dt \;=\; m\,(v_z)_2$$

where x, y and z are the directions of the three axes in three dimensional rectangular co-ordinates and the suffices 1 and 2 refer to the times 't_1' and 't_2', respectively.

Example 5.1

A block of mass 100 kg rests on a smooth horizontal surface as shown in Figure 5.2. If the block is subjected to a constant force of 80 N, inclined at 30° to the horizontal, determine the velocity of the block after 7 seconds. What is the vertical reaction of the horizontal surface on the mass?

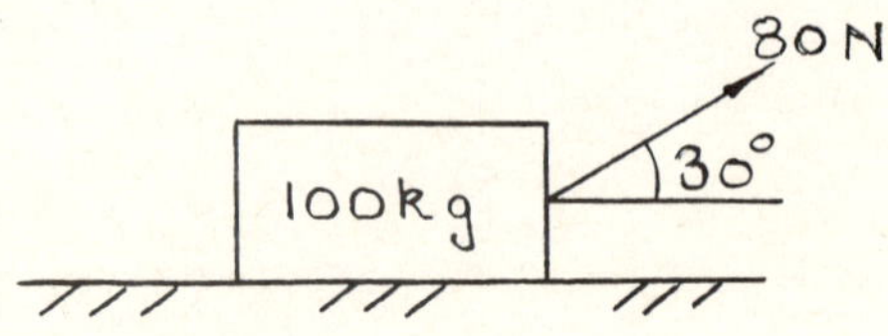

Figure 5.2

The horizontal component of the inclined 80 N force is 80. cos 30° = 69.28 N, and the vertical component of this force is 80. sin 30° = 40 N, acting upwards, as shown by the free body diagram of Figure 5.3.

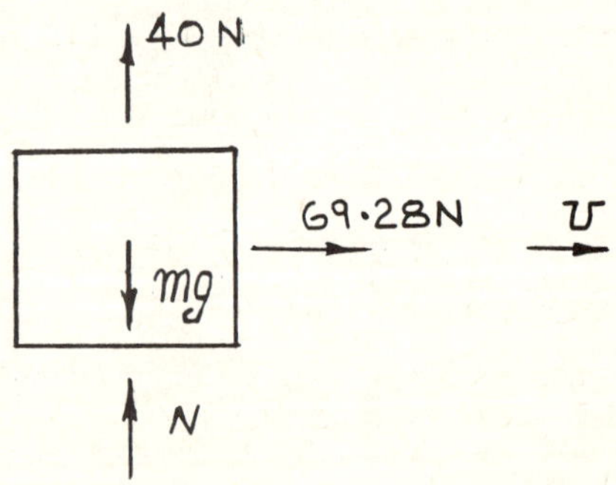

Figure 5.3 Free Body Diagram

From equation (5.6)

$$o + \int F \,.\, dt = mv_2 \text{ in the direction of `}v\text{'},$$

$$\text{or } 69.28 \times 7 = 100 \times v_2$$

$$\text{or } v_2 = 4.85\,\text{m/s}$$

Resolving Vertically

Upward forces = Downward forces

$$N + 40 = mg$$

$$N = 100 \times 9.81 - 40$$

$$\text{or } N = 941 \text{ N}$$

Example 5.2

A block of mass 100 kg is subjected to a force P = 10 t N, where 't' is the time in seconds. If the block's intial velocity, namely 'v_1' is 1 m/s and the coefficient of kinetic friction between the block and the surface it is sliding on, namely, μ_k is 0.25, determine the velocity of the block after 3 seconds.

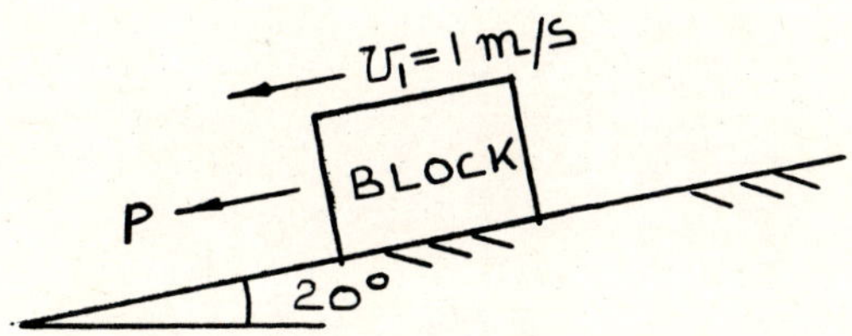

Figure 5.4

The free-body diagram for this block is shown in Figure 5.5.

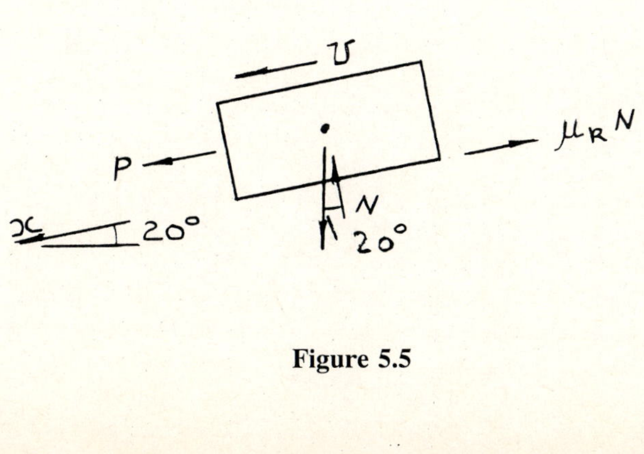

Figure 5.5

Resolving Perpendicular to the Surface

$$N = mg \cos 20°$$

$$= 100 \times 9.81 \times 0.9397$$

$$N = 921.8 \text{ N}$$

Frictional force $= \mu_k N$

Applying equation (5.6) to the free body diagram of Figure 5.5, we get

$$mv_1 + \int_0^3 P \,.\, dt - \mu_k \,.\, N \,.\, t$$

$$+ mg \sin 20° t = mv_2,$$

$$\text{or } 100 \times 1 + \int_0^3 10t \,.\, dt - 0.25 \times 921.8 \times 3$$

$$+ 100 \times 9.81 \times 0.342 \times 3 = 100\, v_2$$

$$\text{or } 100 + 10 \left[t^2/2\right]_o^3 + 691.35 + 1006.5 = 100\, v_2$$

$$\text{or } 100\, v_2 = 460.15,$$

$$\text{or } v_2 = 4.60 \text{m/s}$$

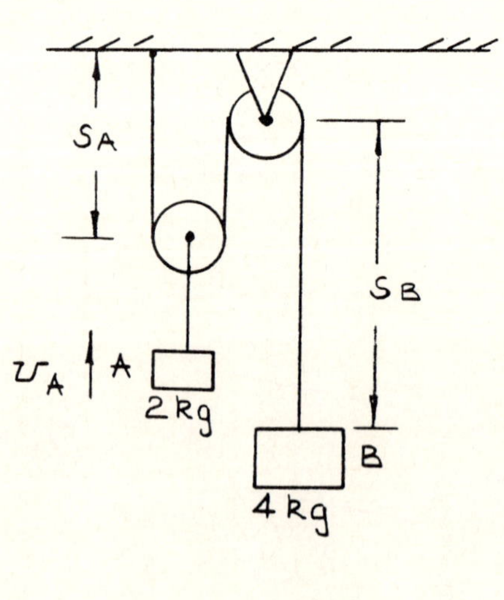

Figure 5.6

Example 5.3

A system of interconnected particles is shown in Figure 5.6. If when $t = 0$, the blocks are at rest, what are the velocities of the blocks after 5s, where 't' is the time in seconds. Neglect the effects of friction and assume that the masses of the cord and pulleys are negligible.

$$2\,S_A + S_B + c = L = \text{length of cord}$$

$$\frac{dL}{dt} = 2\,\dot{S}_A + \dot{S}_B,$$

$$\text{or } \dot{S}_B = -2\,\dot{S}_A,$$

$$\text{but } v_A = -\dot{S}_A$$

$$\text{and } v_B = \dot{S}_B$$

$$\therefore\ v_B = 2\,v_A$$

The free body diagram for particles 'A' and 'B' are shown in Figures 5.7(a) and (b).

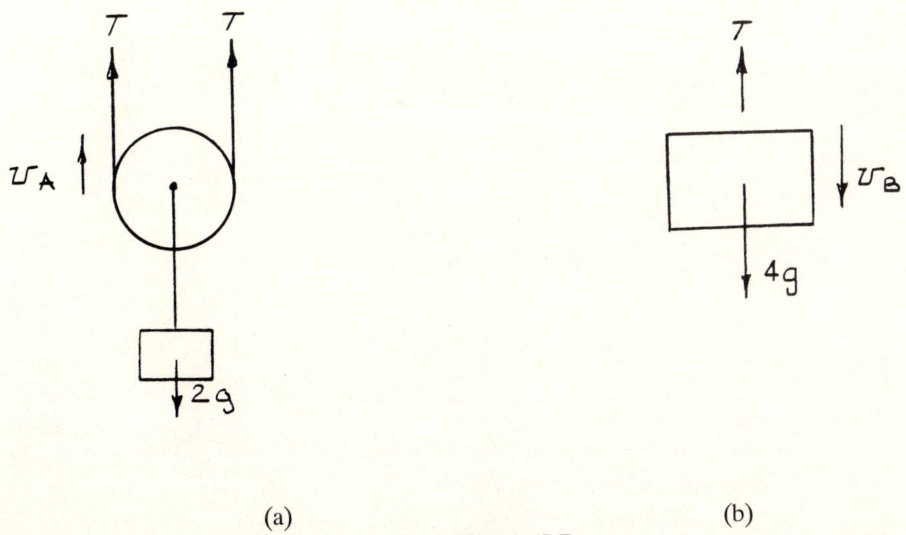

Figure 5.7

Applying equation (5.6) to particle 'A', we get

$$2(v_A)_{t=o} + (-2g + 2T)t = 2(v_A)_{t=5s}$$

$$\text{or } 0 - 19.62 \text{ x } 5 + 2T\,x\,5 = 2\,v_A,$$

$$\text{or } -98.1 + 10T = 2\,v_A,$$

$$\text{or } T = 9.81 + 0.2\,v_A$$

Applying equation (5.6) to particle 'B', we get

$$4(v_B)_{t=0} + (-T + 4g)\,t = 4(v_B)_{t=5\text{s}}$$

$$\text{or } -5T + 196.2 = 4v_B$$

$$\text{or } T = -0.8v_B + 39.24 \qquad (5.9)$$

Equating (5.8) and (5.9), we get

$$9.81 + 0.2v_A = -0.8\,v_B + 39.24,$$

$$\text{but } v_B = 2v_A$$

$$\therefore\; 9.81 + 0.2v_A = -1.6v_A + 39.24,$$

$$\text{or } 1.8v_A = 39.24 - 9.81,$$

$$\text{or } \underline{v_A} = \underline{16.35\,\text{m/s}}$$

$$\text{and } \underline{v_B} = \underline{32.7\,\text{m/s}}$$

From equation (5.8)

$$T = 9.81 + 0.2\,x\,16.35$$

$$\text{or } T = 13.08\,\text{N}$$

Problems 5.1

1. During the firing of a rifle, a constant force '*P*' is applied to a bullet for a period of 0.5s and causes the bullet to reach a speed of 500 m/s. If the mass of the bullet is 2g, what is the magnitude of this impulsive force?

   [2000 N]

2. A hammer head of mass 1 kg strikes a nail of negligible mass at a speed of 15 m/s. If it can be assumed that as a result of this, the nail penetrates a piece of wood and the hammer and nail remain in contact until they come to rest. What is the impulse on the nail?

[15 N.s]

3. A golfer cleanly strikes a golf ball of mass 45 g, which lifts off at an angle of 35° to the horizontal. Assuming that the ground is level and that the air resistance is negligible and also that the golf ball has a range of 50 m, what is the impulsive exerted on the golf ball.

[1.028 N.s]

4. A system of interconnected particles is shown in Figure 5.8. If the system is suddenly released from rest, determine the velocities of the particles after 17s.

[$v_A = v_B = 5.56$ m/s]

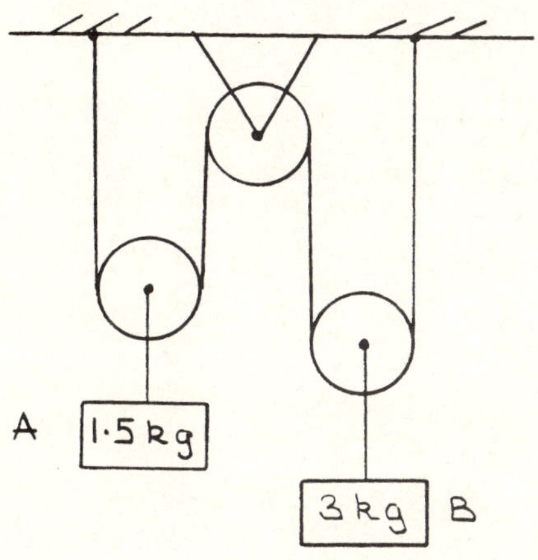

Figure 5.8

5. An aircraft of mass 200 Mg is travelling horizontally with a uniform speed of 80 m/s, when its engines suddenly supply an additional horizontal thrust 'T', where $T = 150 + 3t^2$ kN. Neglecting air resistance and other losses, determine the speed of the aircraft when $t = 4$s.

[83.32 m/s]

6. A tractor of mass 20 Mg is initially at rest. If the tractor is subjected to a force 'F', where $F = 6 - 0.03\,t^2$ kN, determine the speed of the tractor when $t = 5$s. Neglect losses and assume that the tractor travels on a horizontal surface.

   [5.175 km/h]

5.4 CONSERVATION OF LINEAR MOMENTUM

The conservation of linear momentum is useful for problems involving impact; it can be expressed as follows:

linear momentum before impact = *linear momentum after impact*

The conservation of linear momentum will now be applied to a number of problems.

Example 5.4

A locomotive of mass 20 Mg is travelling at a uniform speed of 1.2 m/s when it collides with a stationary wagon of mass 10 Mg. If the two vehicles couple together in 0.5s, determine:

(a) the combined speed of the two vehicles;

(b) the average force between the two vehicles.

(a) Momentum before impact = momentum after impact

$$\text{or } 20 \times 10^3 \times 1.2 = (20 + 10) \times 10^3\, v$$

$$\text{or } v = 20 \times \frac{1.2}{30} = \underline{0.8\,\text{m/s}}$$

(b) Impulse between the two vehicles = momentum

$$\text{or } F \times t = 20 \times 10^3 \times 1.2$$

$$\text{or } F = 20 \times 10^3 \times \frac{1.2}{0.5}$$

$$\underline{F = 48\,\text{kN}}$$

Example 5.5

If the wagon of the previous problem were travelling at 0.4 m/s in the opposite direction to the locomotive, determine:

(a) the common speed of the coupled vehicles;

(b) the average exerted force between the two vehicles.

(a) Momentum before impact = momentum after impact

$$\text{ie } 20 \times 10^3 \times 1.2 - 10 \times 10^3 \times 0.4 = (20 + 10) \times 10^3\, v,$$

$$\text{or } \underline{v = 0.667\,\text{m/s}}$$

(b) Impulse between the two vehicles = momentum

$$\text{or } (20 + 10) \times 10^3 \times 0.667 = F \times 0.5,$$

$$\text{or } \underline{F = 40\ \text{kN}}$$

Example 5.6

A cannon of mass 600 kg fires a projectile of mass 4 kg with a muzzle velocity of 500 m/s in a firing time of 0.02s. Determine:

(a) the recoil velocity of the cannon immediately after firing;

(b) the average impulse force exerted on the projectile.

(a) Momentum before = momentum after

$$4 \times 500 = 600\, v$$

$$\text{or } \underline{v = 3.333\,\text{m/s}}$$

(b) Impulse = momentum

$$\text{or } F \times 0.02 = 4 \times 500$$

$$\text{or } \underline{F = 100\ \text{kN}}$$

Example 5.7

A pile driver of mass 200 kg falls from a height of 1 m and drives a pile of mass 600 kg, vertically into the ground a distance of 0.2 m. Assuming that the pile driver remains in contact with the pile during the motion into the ground, determine the resistance of the ground to motion of the pile, assuming that this resistance is constant.

Motion of Pile Driver

$$v^2 = v_o^2 + 2as$$

$$\text{or } v^2 = 0 + 2 \times 9.81 \times 1$$

$$\underline{v = 4.429\,\text{m/s}}$$

Motion of Pile and Pile Driver

$$\text{Momentum before} = \text{momentum after}$$

$$200 \times 4.429 = (600 + 200)\, v^1$$

$$\therefore \underline{v^1 = 1.107\,\text{m/s}}$$

$$\text{Now } v^2 = v_o^2 + 2as$$

$$\text{or } 0 = 1.107^2 + 2 \times a \times 0.2$$

$$\text{or } \underline{a = -3.065\,\text{m/s}^2}$$

$$\text{From } F = ma$$

$$\text{Resistance of ground} = -800 \times 3.065$$

$$= \underline{-2.452\,\text{kN}}$$

Problems 5.2

1. A locomotive of mass 30 Mg is travelling at 2 m/s when it collides with a wagon of mass 20 Mg travelling in the opposite direction with a speed of 0.5 m/s. If the two vehicles couple together for a time of 0.6s, determine:

   (a) the common velocity of the two vehicles immediately after the collision;

   (b) the average impulsive force exerted between the two vehicles.

   [1 m/s, 83.33 kN]

2. A hammer head of mass 1 kg is travelling vertically downwards with a speed of 25 m/s, when it strikes a nail of mass 0.1 kg, where the latter is placed vertically above a piece of wood. Assuming that the hammer and the nail remain in contact while the nail penetrates the wood a distance of 0.04 m, determine the average resistance of the wood to penetration.

[7.1 kN]

3. A pile driver of mass 150 kg, falls vertically a distance of 1.2m onto a pile of mass 450 kg. If the two remain together while the pile penetrates the ground at a distance of 0.13 m, determine the average resistance of the ground to penetration by the pile.

[2.8 kN]

4. A boy of mass 50 kg walks along a cart of mass 30 kg from one end of the cart to the other, with a uniform speed of 1.5 m/s. Determine the speed that the cart moves and the distance it travels, before he steps off it.

[0.938 m/s 1.25 m]

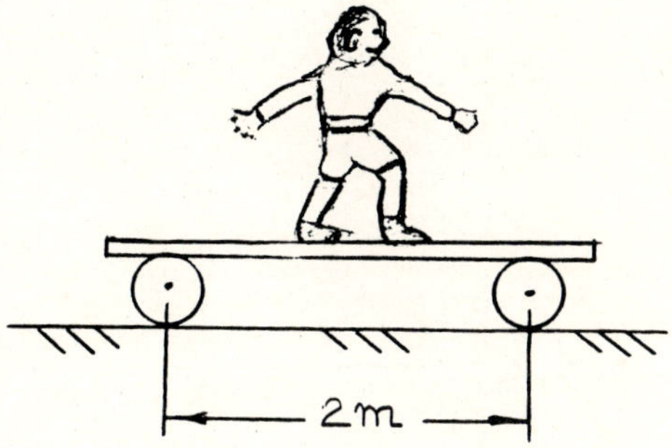

Figure 5.9 Boy on Cart

5. A car of mass 1.5 Mg rest on a barge of mass 15 Mg, in still water, as shown in Figure 5.10. If the car travels the length of the barge at a uniform velocity, determine the distance that the barge moves in the opposite direction.

[4.545 m]

6. A toy car, namely 'A' of mass 0.2 kg has attached to its back a spring of stiffness 400 N/m. Another toy car, namely 'B', of mass 0.3 kg has its back pressed against the spring, so that it compresses the spring by a distance of 0.05 m. If the two cars are suddenly released from rest, determine their velocities when the spring has extended back to its normal length. Neglect frictional effects and the mass of the spring.

[v_A = 7.746 m/s, v_B = 5.164 m/s]

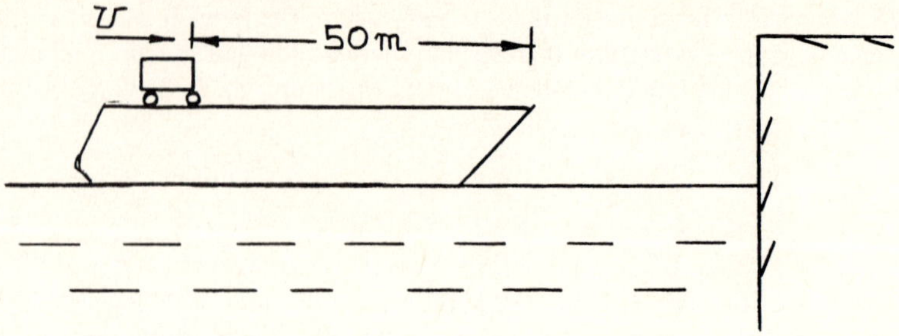

Figure 5.10

5.5 IMPACT

Impact between two particles can result in a relatively large force between the two particles acting for a very short time.

The force between the two particles can cause one particle to increase its velocity and the other particle to decrease its velocity, or alternatively both particles can decrease their velocities or even change their directions of motion.

There are two types of impact, namely *central impact* and *oblique impact*.

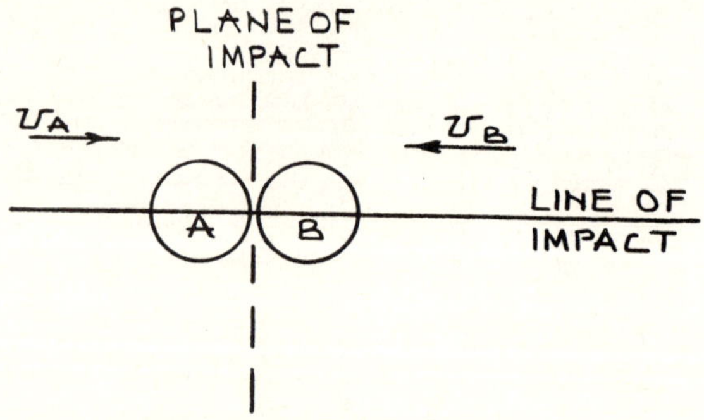

Figure 5.11. Central Impact

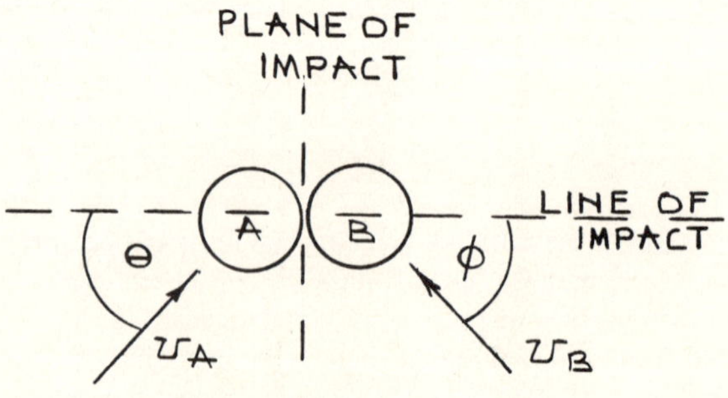

Figure 5.12 Oblique Impact

Central impact is where the *line of impact* is in the direction of motion of the particles, as shown in Figure 5.11 and oblique impact is where the directions of motion are different to the line of impact, as shown in Figure 5.12.
These two types of impact will now be described.

5.5.1 Central Impact

To demonstrate the principles of central impact, consider the two spheres, namely 'A' and 'B' of masses m_A and m_B, respectively. Assume that both masses are travelling in the same direction, as shown in Figure 5.13.

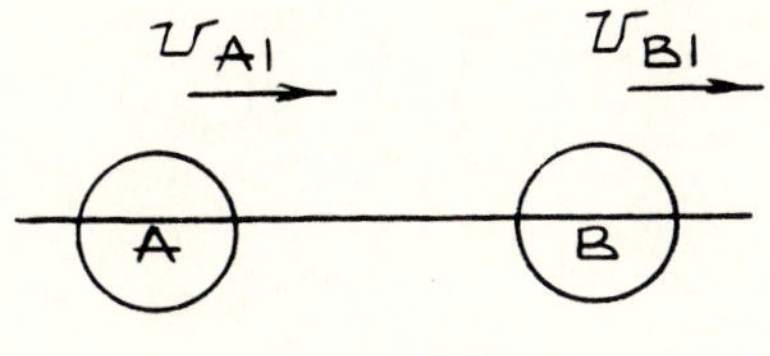

Figure 5.13

It should be assumed that the sphere 'A' is initially travelling with a velocity of v_{A1} and that the sphere 'B' is initially travelling with a velocity v_{B1}, in the same direction as 'A',

$$\text{where } v_{A1} > v_{B1}$$

Eventually the two spheres collide in central impact, as shown in Figure 5.14, where the impact force between the two spheres, namely 'F_d' causes a deformation impulse, namely $\int F_d \cdot dt$.

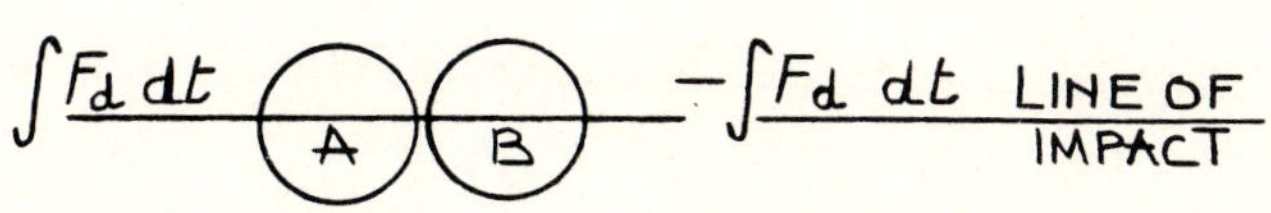

Figure 5.14 Deformation Impulse

In Figure 5.14, the deformation impulse $\int F_d \cdot dt$, is the effect of 'A' on 'B' and the deformation impulse $-\int F_d \cdot dt$, is the effect of 'B' on 'A'; this impulse causes the masses to deform in the manner shown in Figure 5.15, and momentarily travel together with the same velocity v^1.

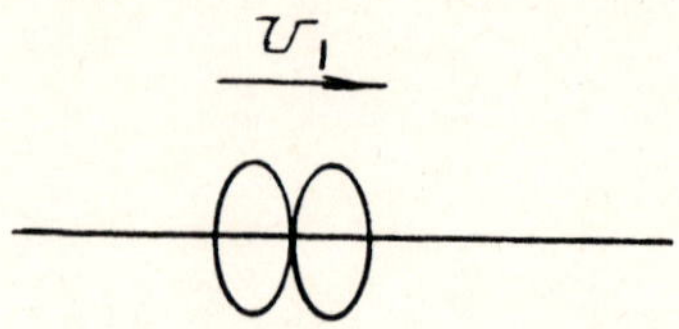

Figure 5.15 Masses Deforming

Now as the masses are elastic, they cause a restitution impulsive force, namely 'F_r', as shown in Figure 5.16.

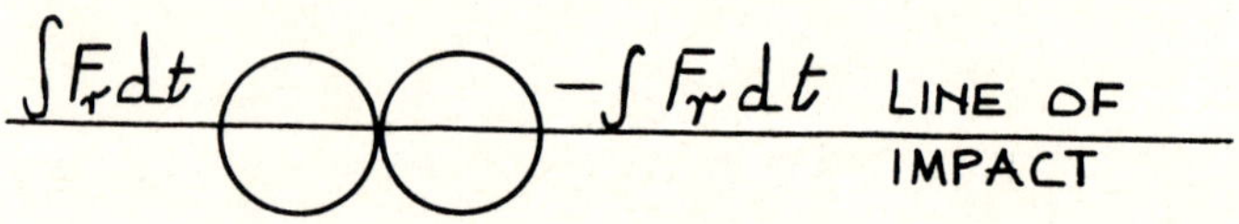

Figure 5.16 Restitution Impulse

The effect of the restitution impulse between the two masses, causes sphere 'A' to decrease its speed and sphere 'B' to increase its speed.

Now the deformation impulse and the restitution impulse are related to the *coefficient of* restitution, namely 'e', according to equation (5.10).

$$e = \frac{\int F_r \cdot dt}{\int F_d \cdot dt} \tag{5.10}$$

Now applying equation (5.6) to the deformation phase for sphere 'A', we get:

$$m_A v_{A1} - \int F_d \cdot dt = m_A v^1, \tag{5.11}$$

$$\text{or} \int F_d \cdot dt = m_A \left(v_{A1} - v^1\right) \tag{5.12}$$

Similarly, applying equation (5.6) to the restitution phase for sphere 'A', we get:

$$m_A\, v^1 - \int F_r \,.\, dt = m_A\, v_{A2}\,, \tag{5.13}$$

$$\text{or} \int F_r \,.\, dt = m_A \left(v^1 - v_{A2}\right) \tag{5.14}$$

where

v_{A2} = the velocity of 'A' after impact.

Substituting equations (5.12) and (5.14) into equation (5.10), we get:

$$e = \frac{m_A\left(v^1 - v_{A2}\right)}{m_A\left(v_{A1} - v^1\right)} = \frac{\left(v^1 - v_{A2}\right)}{\left(v_{A1} - v^1\right)} \tag{5.15}$$

Similarly by applying equation (5.6) to the deformation and restitution phases of sphere 'B', we get:

$$e = \frac{\left(v_{B2} - v^1\right)}{\left(v^1 - v_{B1}\right)} \tag{5.16}$$

where

v_{B2} = the velocity of 'B' after impact.

From equation (5.15),

$$e\left(v_{A1} - v^1\right) = \left(v^1 - v_{A2}\right)$$

$$\text{or } v^1\left(1 + e\right) = e\, v_{A1} + v_{A2} \tag{5.17}$$

Similarly from equation (5.16)

$$e\left(v^1 - v_{B1}\right) = v_{B2} - v^1,$$

$$\text{or } v^1\left(1 + e\right) = v_{B2} + e\, v_{B1} \tag{5.18}$$

Equating (5.17) and (5.18), we get:

$$e\,v_{A1} + v_{A2} = v_{B2} + e\,v_{B1},$$

$$\text{or } e\left(v_{A1} - v_{B1}\right) = v_{B2} - v_{A2},$$

$$\text{or } e = \frac{\left(v_{B2} - v_{A2}\right)}{\left(v_{A1} - v_{B1}\right)} \tag{5.19}$$

$$\text{ie } e = \frac{\text{relative velocity of seperation}}{\text{relative velocity of approach}} \tag{5.20}$$

For perfectly elastic materials, where the particles completely recover their shapes, $e = 1$. For perfectly plastic materials, $e = 0$.

NB: *'e' is larger for an impact of glass on glass than it is for an impact of steel and steel, and for the latter 'e' is larger than for an impact of lead on lead.*

5.5.2 Oblique Central Impact

In this case, consider the two masses namely 'A' and 'B', colliding obliquely, as shown in Figure 5.17.

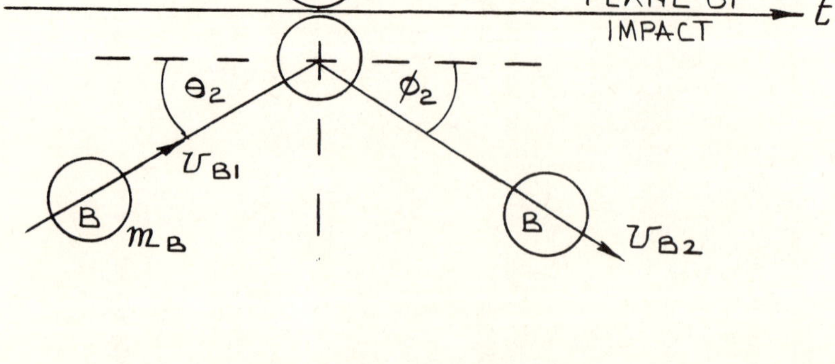

Figure 5.17 Oblique Impact

From Figure 5.17, it can be seen that momentum is conserved in the '*n*' direction, so that

$$m_A (v_{A1})_n + m_B (v_{B1})_n = m_A (v_{A2})_n + m_B (v_{B2})_n \qquad (5.21)$$

Similarly, by considering momentum conservation in the '*t*' direction, we get:

$$m_A (v_{A1})_t = m_A (v_{A2})_t$$

$$\text{and } m_B (v_{B1})_t = m_B (v_{B2})_t$$

$$\text{or } (v_{A1})_t = (v_{A2})_t \qquad (5.22)$$

$$\text{and } (v_{B1})_t = (v_{B2})_t \qquad (5.23)$$

The co-efficient of restitution for oblique impact is given by:

$$e = \frac{(v_{B2})_n - (v_{A2})_n}{(v_{A1})_n - (v_{B1})_n} \qquad (5.24)$$

Once $(v_{A2})_n$ and $(v_{B2})_n$ are known, ϕ_1 and ϕ_2 can be determined.

Example 5.8

A pile driver of mass 500 kg falls vertically through a height of 1.5 m onto a pile of mass 2000 kg. If after impact, the pile driver rises a vertical distance of 0.15 m, determine:

(a) the velocity of the pile immediately after impact;

(b) the coefficient of restitution, '*e*';

(c) the percentage loss of energy due to the first impact.

Let v_{A1} = the velocity of the pile driver before impact.

$$\text{or } v_{A1} = \sqrt{2gh} = \sqrt{2 \times 9.81 \times 1.5} = 5.425 \text{ m/s}$$

Let v_{A2} = the velocity of the pile driver after impact

$$\text{or } v_{A2} = -\sqrt{2gh} = -\sqrt{2 \times 9.81 \times 0.15} = -1.716 \text{ m/s.}$$

(a) Now momentum before = momentum after, so that

$$500 \times 5.425 = -500 \times 1.716 + 2000 \times v_p$$

where v_p = the velocity of the pile immediately after impact.

ie $2712.5 = -858 + 2000\,v_p$

or $v_p = 3570.5/2000$

$= 1.785\,\text{m/s}$

(b)
$$e = \frac{\text{relative velocity after impact}}{\text{relative velocity before impact}}$$

$$= \frac{(v_p - v_{A2})}{(v_{A1} - o)} = \frac{(1.785 + 1.716)}{5.425}$$

(c) KE of pile driver before impact

= PE of pile driver before impact

$= 500 \times 9.81 \times 1.5 = 7357.5\ \text{J}$

KE after impact = KE of pile driver after impact + KE of pile after impact

$$= \frac{500 \times 1.716^2}{2} + \frac{2000 \times 1.785^2}{2}$$

$= 736.16 + 3186.23$

$= 3922.4\ \text{J}$

$$\%\ \text{loss of energy} = \frac{7357.5 - 3922.4}{7357.5} \times 100\%$$

$= 46.7\%$

Example 5.9

Two flat discs, namely disc A and disc B, are travelling on a smooth horizontal surface in the directions shown in Figure 5.18. If the co-efficient of *restitution* $e = 0.7$, determine the magnitudes and directions of the velocities of the discs, immediately after impact.

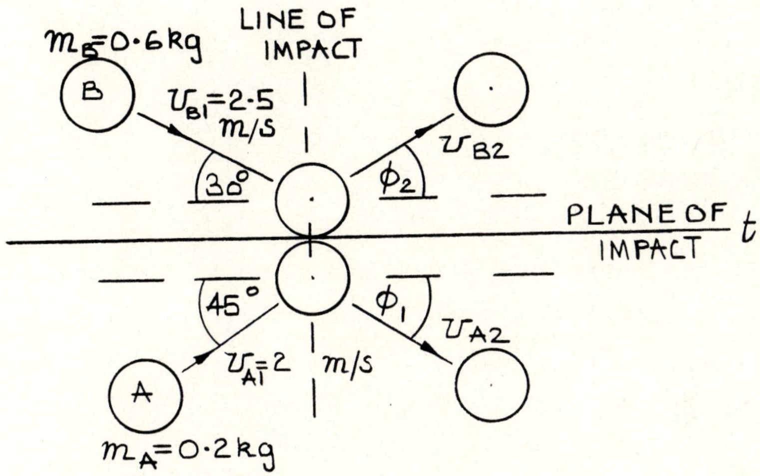

Figure 5.18 Two Discs Colliding

$$(v_{A1})_n = 2 \times \sin 45° = 1.414\,\text{m/s}$$

$$(v_{B1})_n = -2.5 \sin 30° = -1.25\,\text{m/s}$$

$$(v_{A1})_t = 2 \times \cos 45° = 1.414\,\text{m/s}$$

$$(v_{B1})_t = 2.5 \cos 30° = 2.165\,\text{m/s}$$

Considering the conservation of momentum in the '*n*' direction, we get:

$$m_A (v_{A1})_n + m_B (v_{B1})_n = m_A (v_{A2})_n + m_B (v_{B2})_n,$$

$$\text{or } 0.2 \times 1.414 - 0.6 \times 1.25 = 0.2 (v_{A2})_n + 0.6 (v_{B2})_n, \tag{5.25}$$

$$\text{or } (v_{A2})_n + 3 (v_{B2})_n = -2.336$$

From equation (5.20)

$$e = 0.7 = \frac{(v_{B2})_n - (v_{A2})_n}{(v_{A1})_n - (v_{B1})_n}$$

$$= \frac{(v_{B2})_n - (v_{A2})_n}{1.414 - (-1.25)}$$

$$\text{or } 0.7 \times 2.664 = (v_{B2})_n - (v_{A2})_n,$$

$$\text{or } (v_{B2})_n = (v_{A2})_n + 1.8648 \tag{5.26}$$

Substituting equation (5.26) into equation (5.25), we get:

$$(v_{A2})_n + 3\left[(v_{A2})_n + 1.8648\right] = 2.336,$$

$$\text{or } 4(v_{A2})_n = -2.336 - 5.5944$$

$$= -7.9303, \tag{5.27}$$

$$\text{or } (v_{A2})_n = -1.983\,\text{m/s}$$

Substituting equation (5.27) into equation (5.26), we get:

$$(v_{B2})_n = -1.983 + 1.8648,$$

$$\text{or } (v_{B2})_n = -0.118\,\text{m/s}$$

Considering the conservation of momentum in the '*t*' direction, we get:

$$m_A (v_{A1})_t = m_A (v_{A2})_t$$

$$\text{and } m_B (v_{B1})_t = m_B (v_{B2})_t$$

$$\text{or } (v_{A2})_t = (v_{A1})_t = 1.414\,\text{m/s}$$

$$\text{and } (v_{B2})_t = (v_{B1})_t = 2.165\,\text{m/s}.$$

$$\text{Now } \phi_1 = \tan^{-1}\left[(v_{A2})_n \,/\, (v_{A2})_t\right]$$

$$= \tan^{-1}(1.983 \,/\, 1.414),$$

$$\text{or } \phi = 54.51° \text{ (see the diagram below)}$$

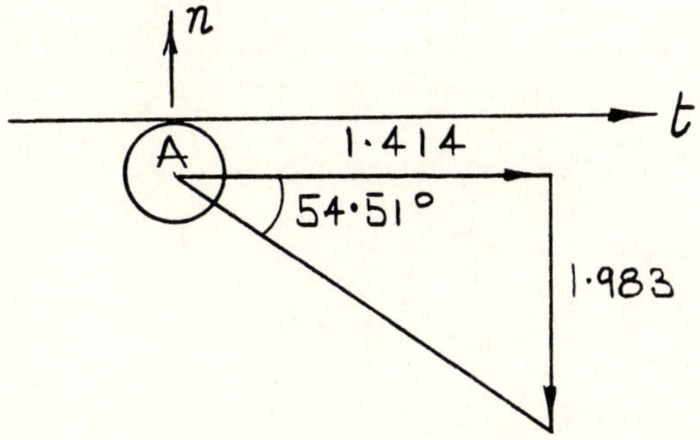

Figure 5.19 Motion of Disc 'A' After Impact

Similarly,

$$\phi_2 = \tan^{-1}\left[(v_{B2})_n \,/\, (v_{B2})_t\right]$$

$$= \tan^{-1}(0.118 \,/\, 2.165) = 3.12° \text{ (see the figure below)}$$

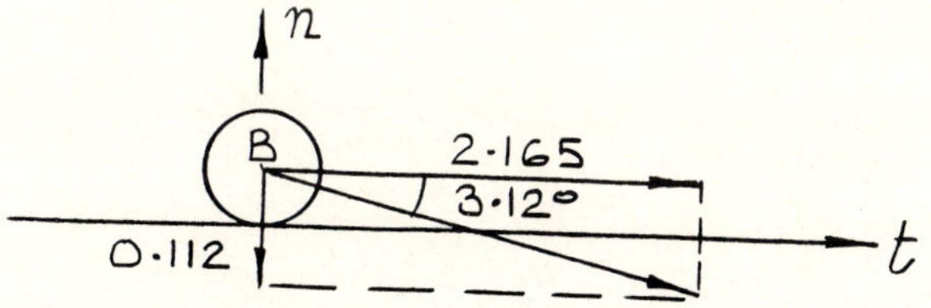

Figure 5.20 Motion of Disc 'B' After Impact

Problems 5.3

1. To test the suitability of a tennis ball for playing a game of tennis, the tennis ball is required to rebound vertically to a height of 1 m, when it is dropped vertically from a height of 1.5 m above the ground. Determine:

(a) the minimum required value of 'e' for the tennis ball;
(b) the percentage loss of energy.

Neglect losses due to air friction and assume that the ground is very firm.

[e = 0.816, 33.3%]

2. A pile driver of mass 1000 kg falls vertically through a distance of 1.8 m on to a pile of mass 2500 kg. If the pile driver rebounds vertically to a height of 0.2 m, determine:

(a) the velocity of the pile immediately after impact;
(b) the value of 'e';
(c) the percentage loss of energy.

[3.17 m/s, e = 0.867, 17.8%]

3. A ball of mass 1 kg hangs vertically downwards on the end of a rubber cord at a distance of 0.8 m from a horizontal ceiling. If the ball is pulled downwards a further 0.2 m, determine how much the rubber cord will stretch again after the ball rebounds. Assume that the stiffness of the rubber cord, namely k = 1000 N/m and e = 0.8.

[0.103 m]

4. A ball is thrown onto a firm horizontal surface at an angle of 40° to the surface, with a velocity of 10 m/s, as shown in Figure 5.21. If e = 0.6, determine the magnitude and direction of the velocity of the ball when it rebounds from the horizontal surface.

[8.576 m/s, 26.73°]

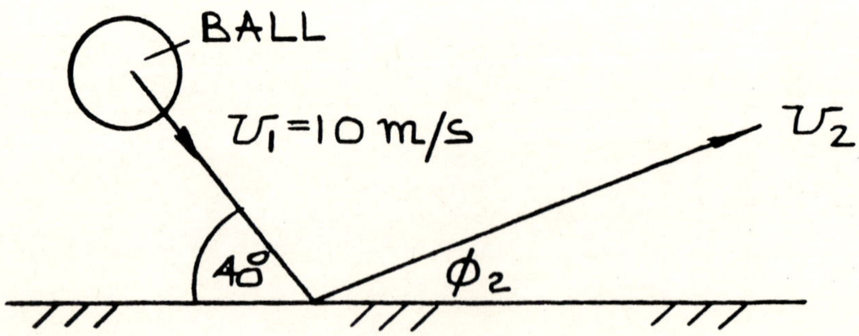

Figure 5.21

5. Two smooth discs moving on a smooth horizontal plane, collide in the manner shown in Figure 5.22. Assuming that $e = 0.8$ between the discs, determine the magnitudes and directions of the resulting velocities after collision.

[$v_{A2} = 0.826$ m/s, $v_{B2} = 1.786$ m/s, $\phi_1 = 51.09°$, $\phi_2 = 34.0°$]

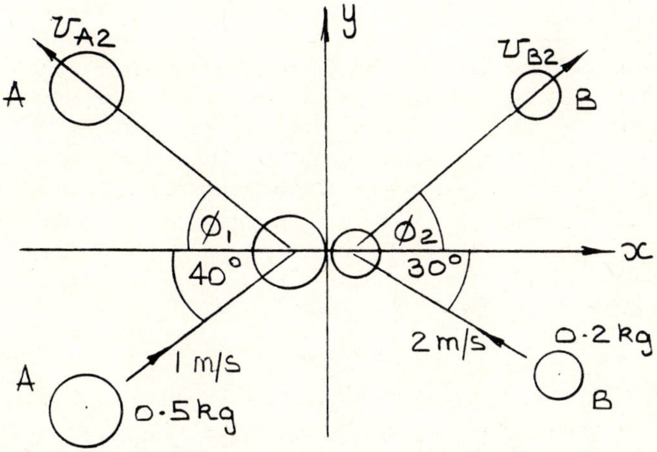

Figure 5.22 Two Discs Colliding

6. Two smooth discs of equal mass lie on a smooth horizontal surface. If disc 'A' travels at a speed of 4 m/s and collides with disc B, which is stationary, in the manner shown in Figure 5.23 determine the magnitudes and directions of the velocities of the discs after collision. Take $e = 0.7$.

[$v_{A2} = 2.066$ m/s, $v_{B2} = 2.944$ m/s, $\phi_A = 14.6°$,

$\phi_B = 90°$ to plane of collision]

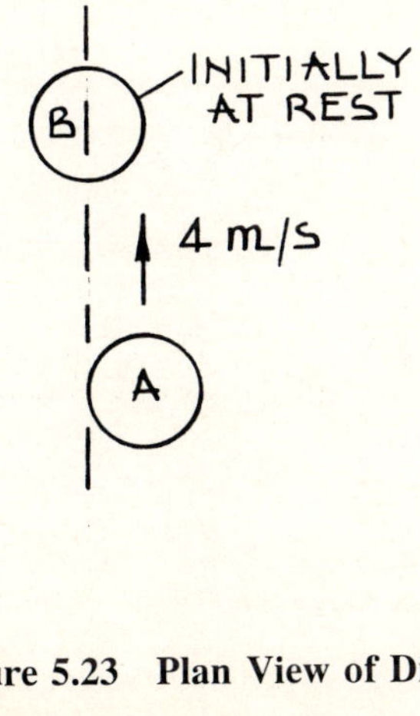

Figure 5.23 Plan View of Discs

5.6 ANGULAR MOMENTUM ($\bar{H}$)

The angular momentum about a point can be described as the moment of momentum about that point.

ie Angular momentum $= \bar{H} = \bar{r} \times m\,\bar{v}$ (5.28)

or the angular momentum is the vector cross product of $\bar{r}$ and $m\bar{v}$.

5.6.1 Consider the Mass m_i, Rotating about the Point 'O'

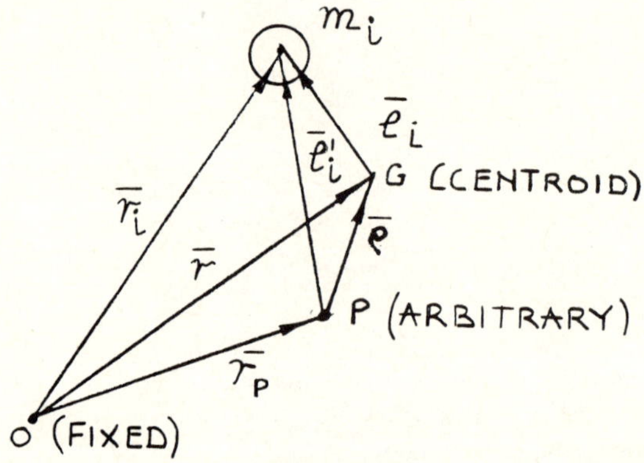

Figure 5.24

The angular momentum $\bar{H}_o$ about the fixed point O, is given by:

$$\bar{H}_o = \sum \bar{r}_i \times m_i\,\bar{v}_i$$

and the time derivative of $\bar{H}_o$, namely, $\dot{\bar{H}}_o$, is given by:

$$\dot{\bar{H}}_o = \sum \dot{\bar{r}}_i \times m_i\,\bar{v}_i + \sum \bar{r}_i \times m_i\,\dot{\bar{v}}_i, \qquad (5.29)$$

but $\dot{\bar{r}}_i$ and $\bar{v}_i$ are identical vectors so that $\dot{\bar{r}}_i \times \bar{v}_i = 0$

Hence, equation (5.29) becomes

$$\dot{\bar{H}}_o = \sum \bar{r}_i \times m_i\,\dot{\bar{v}}_i \qquad (5.30)$$

but $\dot{\bar{v}}_i = \bar{a}_i$ = the acceleration vector for the mass m_i

$$\therefore \dot{\bar{H}}_o = \sum \bar{r}_i \times m_i \, \bar{a}_i$$

$$\text{but} \sum \bar{F} = \sum m_i \, \bar{a}_i \qquad (5.31)$$

$$\therefore \dot{\bar{H}}_o = \sum \bar{r}_i \times \bar{F}_i$$

$$= \sum \bar{M}_o$$

where

$\Sigma \bar{M}_o$ = the sum of the moments of all the accelerating forces about 'O'

5.5.2 Expression for the Time Derivative of the Angular Momentum about 'G'

Now H_G = angular momentum of the mass m_i about the centre of gravity 'G'

$$\bar{H}_G = \sum \bar{\rho}_i \times m_i \, \dot{\bar{r}}_i$$

$$\text{and } \dot{\bar{H}}_G = \sum \bar{\rho}_i \times m_i \, \ddot{\bar{r}}_i + \sum \dot{\bar{\rho}}_i \times m_i \, \dot{\bar{r}}_i$$

$$\text{but } \dot{\bar{r}}_i = \dot{\bar{r}} + \dot{\bar{\rho}}_i$$

$$\therefore \dot{\bar{H}}_G = \sum \bar{\rho}_i \times m_i \, \ddot{\bar{r}}_i + \sum \dot{\bar{\rho}}_i \times m_i \left(\dot{\bar{r}} + \dot{\bar{\rho}}_i\right), \qquad (5.32)$$

$$\text{but } \dot{\bar{r}} \times \sum m_i \, \dot{\bar{\rho}}_i = \dot{\bar{r}} \times \frac{d}{dt} \sum m_i \, \bar{\rho}_i \qquad (5.33)$$

Equation (5.33) vanishes, since

$\Sigma \, m_i \, \rho_i$ = 0 (ie the moment of the mass about the centroid is zero)

Also $\Sigma \, \dot{\bar{\rho}}_i \times \dot{\bar{\rho}}_i$ = 0, since the cross product of two identical vectors is zero.

Hence, equation (5.32) becomes:

$$\dot{\bar{H}}_G = \sum \bar{\rho}_i \times m_i \, \ddot{\bar{r}}_i \qquad (5.34)$$

$$\text{but} \sum m_i \, \ddot{\bar{r}}_i = \sum \bar{F}_i$$

$$\therefore \dot{\bar{H}}_G = \sum \bar{\rho}_i \times \bar{F}_i \qquad (5.35)$$

$$= \sum \bar{M}_G$$

where

$$\sum F_i = \sum m_i \bar{\ddot{r}}_i$$

5.5.3 Expression for the Time Derivative of the Angular Momentum about the Point 'P'

$\bar{H}_p$ = angular momentum about the point 'P'

$$\bar{H}_p = \sum \bar{\rho}_i^{\prime} \times m_i \bar{\dot{r}}_i$$

$$\text{but } \bar{\rho}_i^{\prime} = \bar{\rho} + \bar{\rho}_i$$

$$\therefore \bar{H}_p = \sum \left(\bar{\rho}_i + \bar{\rho}_i\right) \times m_i \bar{\dot{r}}_i$$

$$\text{and } \bar{\rho} \times \sum m_i \bar{v}_i = \bar{\rho} \times m \bar{v}$$

$$\text{Now, } \sum \bar{\rho}_i \times m_i \bar{v}_i = H_G$$

$$\therefore \bar{H}_p = \bar{H}_G + \bar{\rho} \times m \bar{v} \tag{5.36}$$

Similarly

$$\sum \bar{M}_p = \sum \bar{M}_G + \bar{\rho} \times m \bar{a} \tag{5.37}$$

where

$\Sigma \bar{M}_p$ = moment of the accelerating forces about the point 'P'

5.7 CONSERVATION OF ANGULAR MOMENTUM

The law of the conservation of angular momentum states that "the angular momentum before impact is equal to the angular momentum after impact".

5.7.1 Angular Impulse

The angular impulse of a couple about a point 'O' is given by the expression:

$$\sum \int_{t_1}^{t_2} \bar{M}_o \, . \, dt = \left(\bar{H}_o\right)_2 - \left(\bar{H}_o\right)_1 \tag{5.38}$$

where

$$\sum \int_{t_1}^{t_2} \overline{M}_o \, . \, dt \;=\; \text{angular impulse about `O'}$$

$$\left(\overline{H}_o\right)_1 \;=\; \text{initial angular momentum about `O'}$$

$$\left(\overline{H}_o\right)_2 \;=\; \text{final angular momentum about `O'}$$

In general, the conservation of angular momentum about 'O' can be expressed by equation (5.39).

$$\left(\overline{H}_o\right)_1 + \sum \int_{t_1}^{t_2} \overline{M}_o \, . \, dt = \left(\overline{H}_o\right)_2 \tag{5.39}$$

Equation (5.39) can be seen to be the angular version of equation (5.6)

Example 5.10

Two metal balls, each of mass '*m*', are attached to the ends of a rod of negligible mass, as shown in Figure 5.25. If this combination is resting on a smooth horizontal surface and it is subjected to a suddenly applied force '*F*', determine:

(a) the instantaneous acceleration of the mass centre;

(b) the angular acceleration $\ddot{\theta}$ about the centre of gravity 'G'.

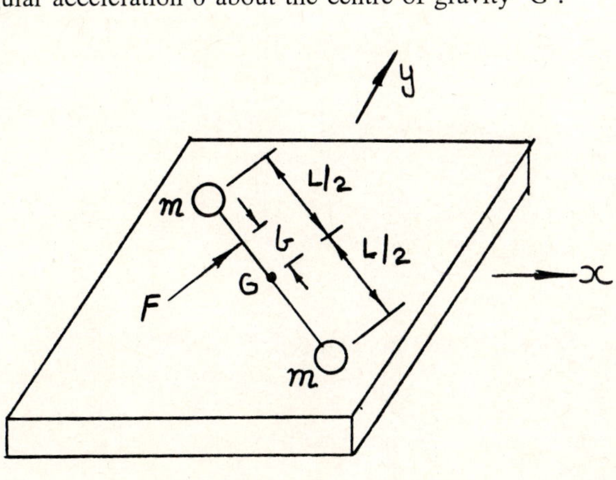

Figure 5.25

(a) Now force = mass x acceleration

$$\text{or } F = 2\,m\,\overline{a}$$

$$\therefore \overline{a} = \frac{F}{(2m)}$$

(b)

$$\dot{\overline{H_G}} = \sum \overline{M_G}$$

$$\text{and } \overline{H_G} = \text{moment of momentum}$$

$$= 2\,m\,.\,\frac{L}{2}\,.\,\dot{\theta}\,\frac{L}{2},$$

$$\text{or } \overline{H_G} = m\,L^2\,\frac{\dot{\theta}}{2}$$

$$\text{where } \dot{\theta} = \text{angular velocity}$$

$$\therefore \dot{\overline{H_G}} = m\,L^2\,\frac{\ddot{\theta}}{2} = \sum M_G,$$

$$\text{where } \ddot{\theta} = \text{angular acceleration}$$

$$\text{Now } \sum M_G = Fb$$

$$\therefore Fb = m\,L^2\,\frac{\ddot{\theta}}{2}$$

$$\text{or } \ddot{\theta} = \frac{2\,Fb}{(m\,L^2)}$$

Example 5.11

A particle of mass '*m*' travels in the arc of a circle of radius '*r*' with a speed of '*v*'. Determine:

(a) the angular momentum about the point 'O';

(b) the rate of increase of the speed at any angle 'θ'.

Neglect frictional effects.

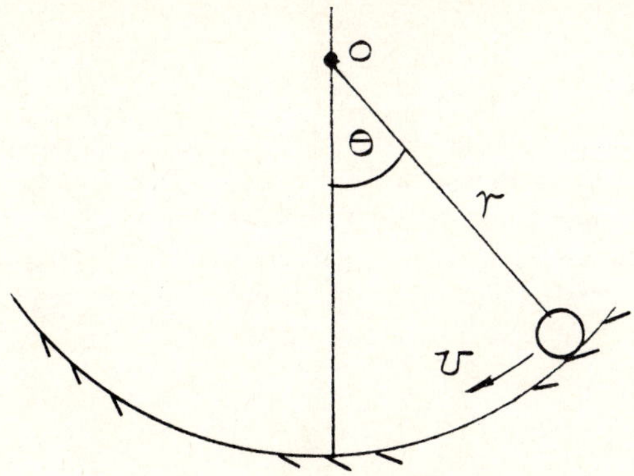

Figure 5.26

(a) Angular momentum about 'O' = $\overline{H}_o = r \times mv$ (5.40)

(b) From the free body diagram of Figure 5.27

$$\sum \overline{M}_o = mg \, . \, \sin\theta \, . \, r \tag{5.41}$$

Now $\dot{\overline{H}}_o = \dfrac{\mathrm{d}\overline{H}_o}{\mathrm{d}t}$,

Hence from equation (5.40)

$$\dot{\overline{H}}_o = mr \, . \, \dot{v} \tag{5.42}$$

$$\text{but } \dot{\overline{H}}_o = \sum \overline{M}_o$$

$$\therefore mr \, . \, \dot{v} = mg \, . \, \sin\theta \, . \, r$$

$$\text{or } \dot{v} = \frac{\mathrm{d}v}{\mathrm{d}t} = g \sin\theta$$

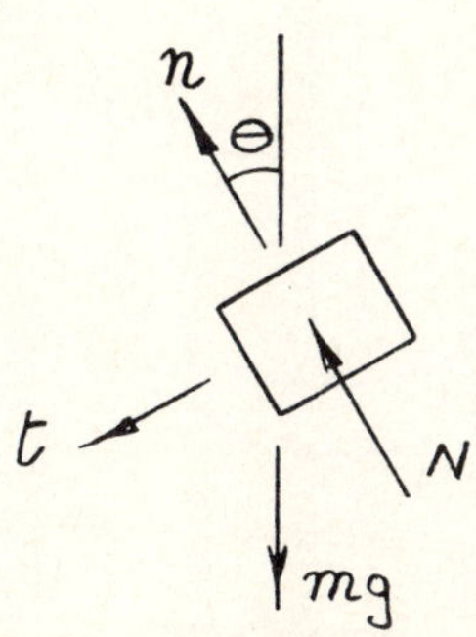

Figure 5.27

Problems 5.4

1. A particle of mass 3 kg is designed to rotate about the point 'O' on a smooth horizontal surface, as shown in Figure 5.28. The particle is attached to a rod of negligible mass, and the system is subjected to a couple of $2t$ N.m, where 't' is the time in seconds. Additionally, the particle is subjected to a horizontal force of 5 N, acting in the circumferential direction. If the particle is initially at rest, determine the speed of the particle after a period of 4.5s.

   [23.61 m/s]

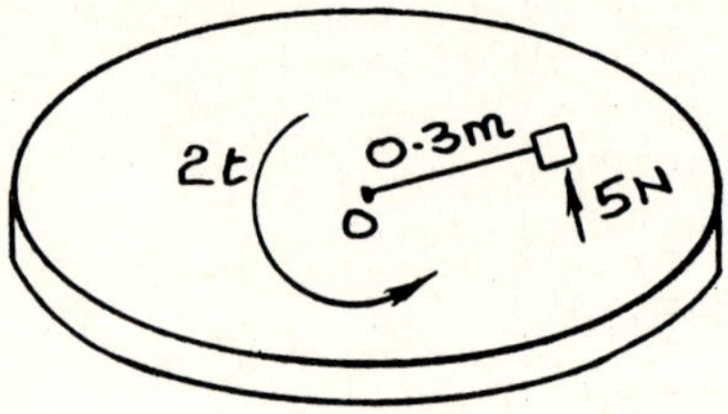

Figure 5.28

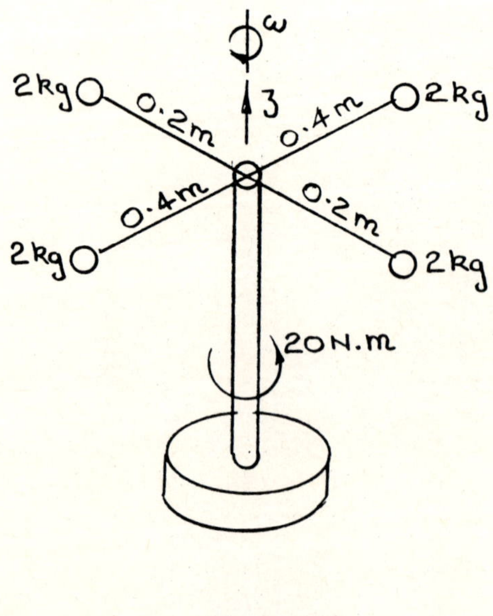

Figure 5.29

2. An artefact consists of four masses which are rigidly attached to a vertical shaft by four horizontal spokes of negligible mass, as shown by Figure 5.29. The system is initially rotating clockwise with a constant angular velocity of 15 rads/s, when looking in a downward direction. If the shaft is suddenly subjected to a braking torque of 20 N.m, determine how long it will take for the system to come to rest?

[0.6s]

3. Four 3 kg masses are rigidly attached to four light horizontal crossbars which are in turn, rigidly attached to a light vertical spindle which is free to rotate about the 'z' axis. If starting from rest the spindle is subjected to a couple 'M_o' of magnitude $2.5t + 1.5$ N.m, where 't' is the time in seconds, determine the speed of the particles after a period of 3.5s. Neglect looses due to friction.

[1.14 m/s]

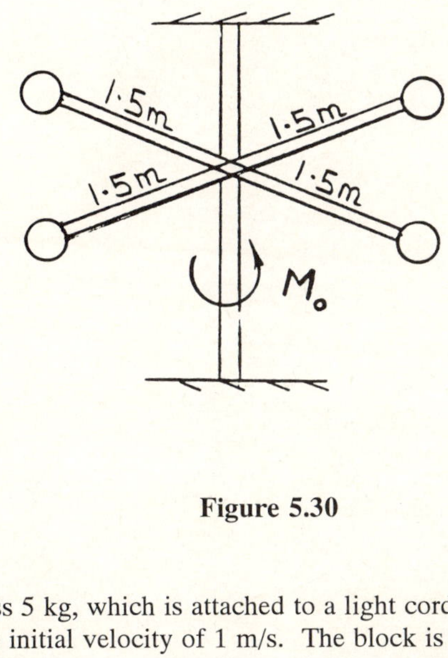

Figure 5.30

4. A block of mass 5 kg, which is attached to a light cord, rotates on a rough horizontal surface with an initial velocity of 1 m/s. The block is then subjected to two constant forces, one of these being a radial force of 8 N and the other being a horizontal force of 30 N acting at 30° to the circumference of the table, as shown in Figure 5.31. If the coefficient of kinetic friction between the block and the surface of the table is 0.4, determine the time taken for the tension in the cord to reach 80 N.

[2.46s]

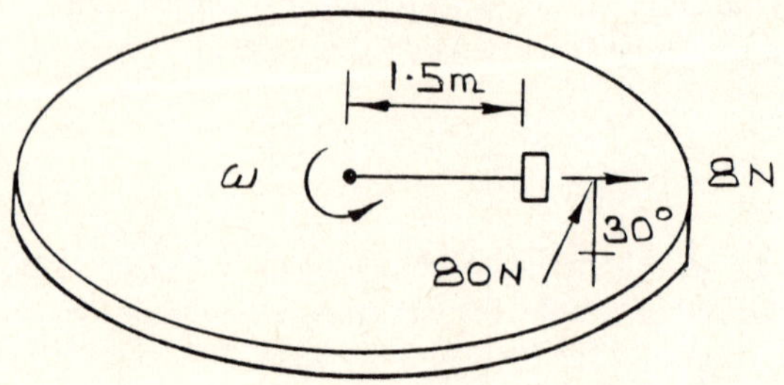

Figure 5.31

CHAPTER SIX

Kinematics of a Rigid Body

6.1 INTRODUCTION

A rigid body is a body that is assumed not to change its shape. The kinematics of a rigid body is the study of the motion of the body without reference to the forces causing the motion. Unlike the kinematics of particles, the kinematics of a rigid body usually involves, both the translation and rotation of the rigid body.

The kinematics of a rigid body is of much importance in a number of different branches of engineering, including in the design of cams and gears and in the motion of machine components.

6.2 RIGID BODY MOTION

In planar motion, there are three types of rigid body motion; these are:

(i) translation of a rigid body;
(ii) rotation of a rigid body;
(iii) general plane motion of a rigid body involving both translation and rotation.

6.2.1 Translation of a Rigid Body

There are two types of translatory rigid body motion, namely, rectilinear motion, as shown in Figure 6.1(i) and curvilinear motion, as shown in Figure 6.1(ii).

(i) Rectilinear Motion **(ii) Curvilinear Motion**

Figure 6.1 Translatory Motion of a Rigid Body

In both Figures 6.1(i) and 6.1(ii), the rigid body does not rotate, but remains parallel throughout its motion.

6.2.2 Motion about a Fixed Axis

In this case, the rigid body rotates about a fixed axis, as shown in Figure 6.2. The rigid body does not translate.

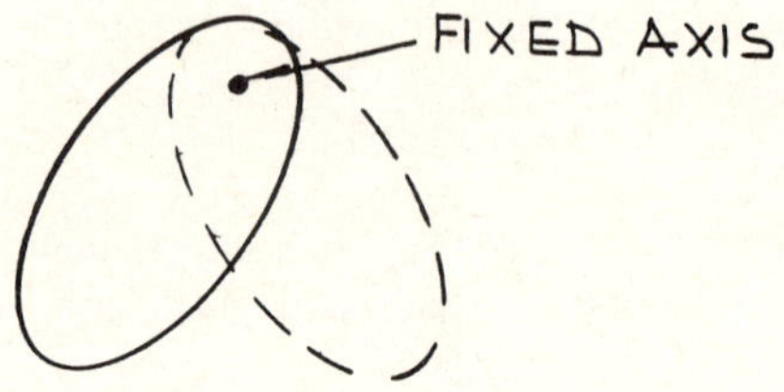

Figure 6.2 Rotation about a Fixed Point

6.2.3 General Planar Motion

In this case, the rigid body rotates and translates during motion, as shown in Figure 6.3.

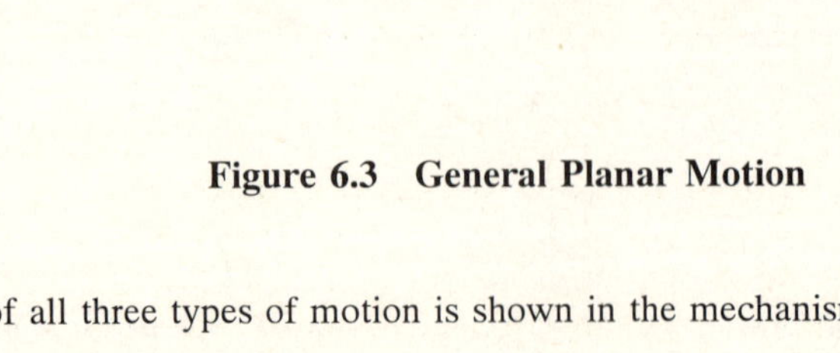

Figure 6.3 General Planar Motion

An example of all three types of motion is shown in the mechanism of Figure 6.4.

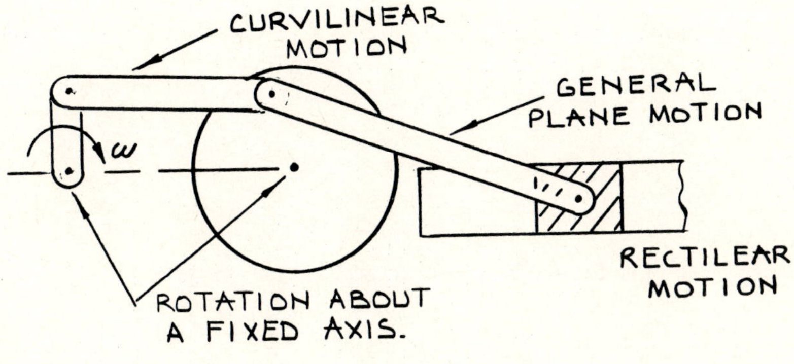

Figure 6.4

6.3 TRANSLATION

Consider translatory motion of the rigid body of Figure 6.5, with reference to the points A and B.

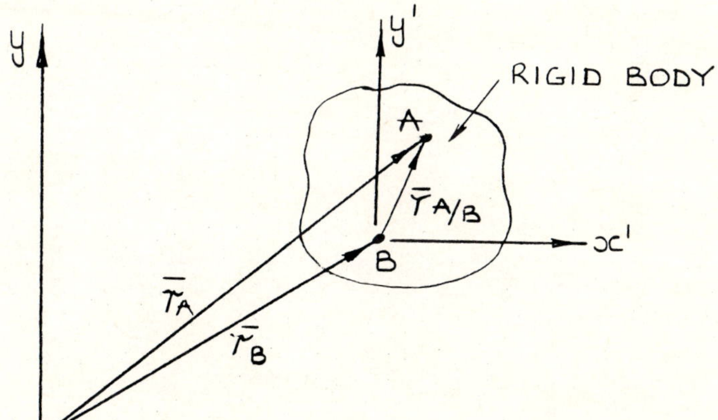

Figure 6.5 Translatory Motion of a Rigid Body

The x-y co-ordinate system is the fixed or reference system and the x^1-y^1 co-ordinate system has its origin at any point 'B' on the rigid body, where the x^1 - y^1 system is parallel to the x-y system.

Let

$\overline{r_A}$ = the position vector of the point `A′

$\overline{r_B}$ = the position vector of the point `B′

$\overline{r_{A/B}}$ = the relative position of `A′ with respect to B the $x^1 - y^1$ origin, so that

$$\overline{r_A} = \overline{r_B} + \overline{r_{A/B}} \tag{6.1}$$

Now if equation (6.1) is differentiated with respect to time, we get

$$\overline{\dot{r}_A} = \overline{\dot{r}_B} + \overline{\dot{r}_{A/B}} \tag{6.2}$$

but $\bar{\dot{r}}_{A/B} = 0$, because $\bar{r}_{A/B}$ is a constant, as the body is rigid and not flexible.

$$\therefore \overline{\dot{r}_{A/B}} = 0$$

$$\text{and } \overline{\dot{r}_A} = \overline{\dot{r}_B} \tag{6.3}$$

$$\text{or } \overline{v_A} = \overline{v_B}$$

Similarly

$$\overline{\dot{v}_A} = \overline{\dot{v}_B}$$

$$\text{or } \overline{a_A} = \overline{a_B} \tag{6.4}$$

Hence, from equations (6.3) and (6.4), it can be concluded that for the rectilinear and curvilinear motion of a rigid body, all points move with the same velocity and acceleration. Thus, the methods used in the kinematics of a particle apply equally as well as in the translatory kinematics of a rigid body.

NB An alternative form for equation 6.1 is:

$$\overline{r_B} = \overline{r_A} + \overline{r_{B/A}}, \tag{6.5}$$

where $\bar{r}_{B/A}$ = the relative position vector of 'B' with respect to 'A'.

6.4 ROTATION ABOUT A FIXED AXIS

An example of the rotation of a rigid body about a fixed axis is given in Figure 6.6.

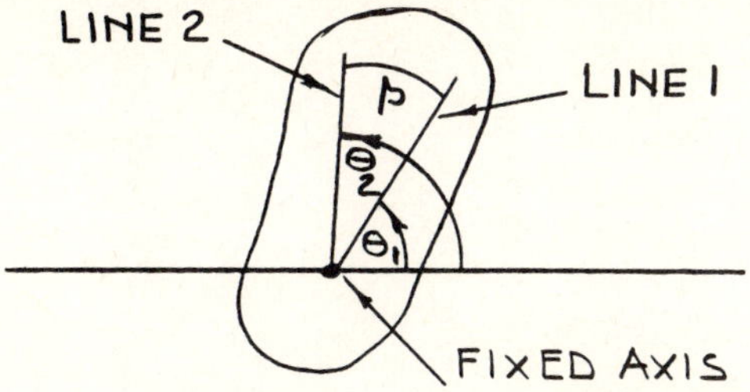

Figure 6.6 Rotation of a Rigid Body

Consider the rotation of a rigid body about a fixed axis, as shown in Figure 6.6. Consider also the motion of two straight lines on this rigid body. Let line 1 lie at an angle θ_1 to the horizontal and let line 2 lie at an angle, θ_2 to the horizontal, so that the angle between the two lines is β.

$$\text{Now } \theta_2 = \theta_1 + \beta$$
$$\text{and } \dot{\theta}_2 = \dot{\theta}_1 \qquad (6.6)$$

It should be noted that $\dot{\beta} = 0$, because β is a constant.

$$\text{Hence, } \ddot{\theta}_2 = \ddot{\theta}_1 \qquad (6.7)$$

$$\text{Similarly, } \Delta\theta_2 = \Delta\theta_1 \qquad (6.8)$$

Hence, from equations (6.6) to (6.8), it can be concluded that in the rotation of a rigid body, all lines have the same angular displacement, angular velocity and angular acceleration.

It should be noted that the angular motion need not be about a fixed axis normal to the plane of motion.

6.2.1 Angular Motion Relationships

The equations for angular motion are:

$$\text{Angular velocity} = \omega = \frac{d\theta}{dt} \qquad (6.9)$$

$$\text{Angular acceleration} = \alpha = \frac{d\omega}{dt} = \frac{d\omega}{d\theta} \cdot \frac{d\theta}{dt} \qquad (6.10)$$

$$= \omega \cdot \frac{d\omega}{d\theta} \tag{6.11}$$

$$\text{or } \alpha \, d\theta = \omega \cdot d\omega \tag{6.12}$$

$$\text{or } \ddot{\theta} \cdot d\theta = \dot{\theta} \cdot d\dot{\theta} \tag{6.13}$$

In cases where the **angular acceleration** is constant, the following equations can be readily derived:

$$\omega = \omega_0 + \alpha t \tag{6.14}$$

$$\omega^2 = \omega_0^2 + 2 \alpha (\theta_0 - \theta) \tag{6.15}$$

$$\theta = \theta_0 + \omega_0 t + \frac{1}{2} \alpha t^2 \tag{6.16}$$

where

θ_0 = initial angular displacement

ω_0 = initial angular velocity

t = time

It should be noted that equations (6.14) to (6.16) are the angular equivalents of equations (2.8), (2.9) and (2.12).

6.4.2 Rotation about a Fixed Axis

Consider the rotation of the rigid body about a fixed axis '0', as shown in Figure 6.7.

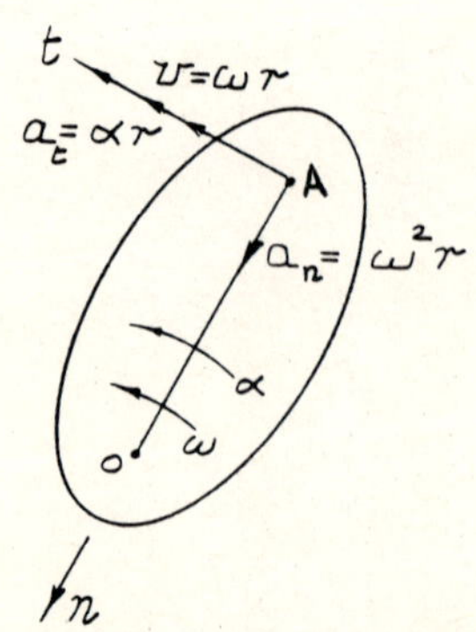

Figure 6.7 Rotation about a Fixed Axis

Consider the motion of the point 'A' on the rigid body. The normal 'n' and tangential 't' axes are shown. Additionally, ω and α are shown positive.

At the point 'A

v = $\omega\, r$

a_n = acceleration in the 'n' direction

= $\omega^2 r = v^2 / r = v\omega$

a_t = acceleration in the 't' direction

= $\alpha\, r$

In vector form, the above expressions can be put in the following form:

$$\bar{v} = \dot{\bar{r}} = \bar{\omega} \times \bar{r} \qquad (6.17)$$

$$\bar{a} = \dot{\bar{v}} = \bar{\omega} \times \dot{\bar{r}} + \dot{\bar{\omega}} \times \bar{r}$$

$$= \bar{\omega} \times (\bar{\omega} \times \bar{r}) + \dot{\bar{\omega}} \times r$$

$$= \bar{\omega} \times (\bar{\omega} \times \bar{r}) + \bar{\alpha} \times \bar{r}$$

but, $\bar{a} = \bar{a}_n + \bar{a}_t$,

so that

$$\bar{a}_n = \bar{\omega} \times (\bar{\omega} \times \bar{r})$$

and $\bar{a}_t = \bar{\alpha} \times \bar{r}$

It should be noted that

$$\bar{r} \times \bar{\omega} = -\bar{v}$$

Example 6.1

A wheel which has a cord wrapped around its external circumference is initially at rest but free to rotate about the point '0', as shown in Figure 6.8.

If this cord is given a sudden pull, so that the wheel accelerates with a value of $a = 2t^2$ m/s^2, where 't' is the time in seconds, determine in terms of 't':

(i) the velocity of 'P';

(ii) the angular position of the line OP.

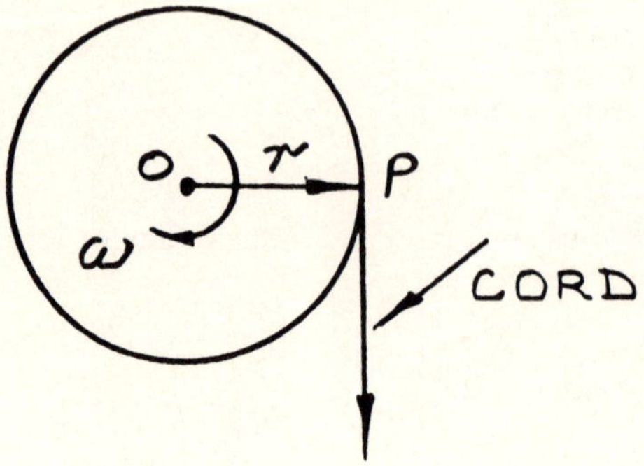

Figure 6.8

(a) $$a = 2t^2 = \alpha r \text{ m/s}^2,$$

$$\text{or } \alpha = 2t^2/r \text{ rad/s}^2,$$

$$\text{but } \alpha = \frac{d\omega}{dt}$$

$$\therefore \frac{d\omega}{dt} = 2t^2/r$$

$$\text{and } d\omega = (2t^2/r)\, dt$$

$$\text{or } \int_o^\omega d\omega = \int_o^t (2t^2/r)\, dt$$

$$\text{or } \omega = 2t^3/(3r)$$

$$\text{Now } v = \omega r$$

$$\therefore v = 2t^3/3 \text{ m/s}$$

(b) Now $\omega = \dfrac{d\theta}{dt}$

or $\dfrac{d\theta}{dt} = 2t^3/(3r)$

or $\displaystyle\int_o^\theta d\theta = \int_o^t 2t^3/(3r)\ .\ dt$

or $\theta = 2t^4/(12r)$

or $\theta = t^4/(6r)$ *rads*

Example 6.2

The two disks, namely disk 'A' and disk 'B' are initially at rest. If disk 'A' is made to rotate with a constant angular acceleration of $\alpha_A = 1.5$ rad/s^2, and this causes disk 'B' to rotate without slipping, determine the angular velocity and angular acceleration of disk 'B' after 5 revolutions.

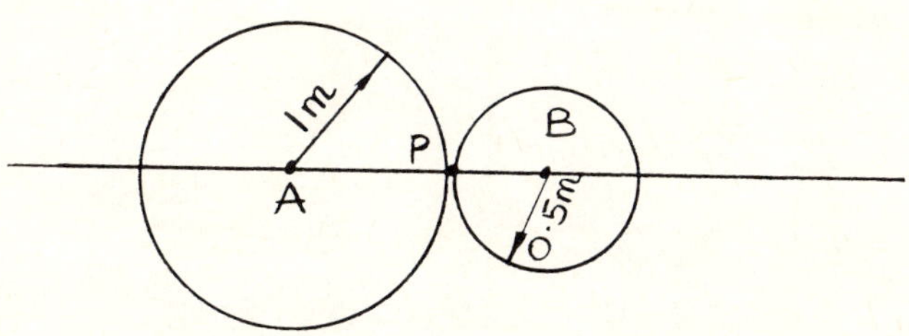

Figure 6.9

Consider point '*P*'

Acceleration of disk 'A' at the point '*P*' $= \quad a_A = \alpha_A \times 1 = 1.5$ m/s^2

but acceleration of disk 'B' at the point '*P*' $= \quad a_B = a_A = 1.5$ m/s^2

$\therefore$ $\boldsymbol{\alpha_B = 1.5/0.5 = 3}$ **rad/s^2**

α_A and α_B are constants, hence equation (6.15) can be used as follows:

$$\omega^2 = \omega_o^2 + 2\alpha(\theta - \theta_o)$$

After 5 revolutions

$$\omega_A^2 = 0 + 2 \times 1.5 \times 2\pi \times 5 = 94.248$$

$$\text{or } \omega_A = 9.708 \text{ rad/s}$$

$$v_A = \omega_A \times 1m = 9.708 \text{ m/s} = v_B$$

$$\therefore \omega_B = v_B/0.5 = 19.416 \text{ rad/s}$$

Example 6.3

The right angled bar rotates in a clockwise direction about the fixed point 'O'. The bar is rotating with an angular velocity of 3 rad/s, when it suddenly decelerates with a constant deceleration of 5 rad/s^2. Determine the velocity and acceleration of the point 'A'.

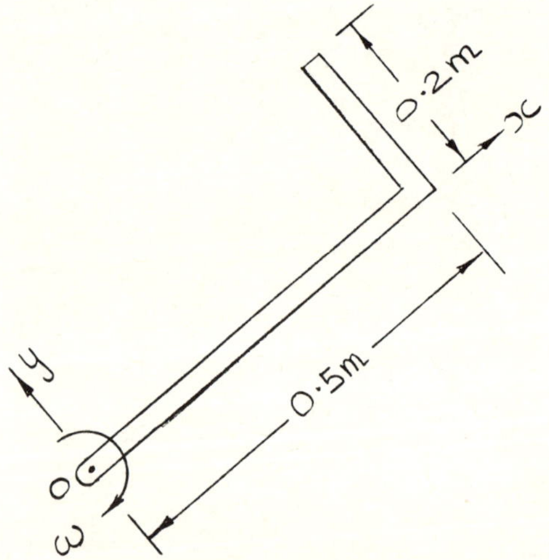

Figure 6.10

As the 'z' axis is perpendicularly out of the paper, 'ω' is negative according to the right hand screw rule and 'α' is positive.

$$\therefore \omega = -3k \ rad/s$$

$$and\ \alpha = 5k \ rads/s^2$$

$$\bar{r} = 0.5\,\bar{i} + 0.2\,\bar{j}$$

$$\text{Now } \overline{v} = \overline{\omega} \times \overline{r}$$

$$= -3k \times (0.5\,\overline{i} + 0.2\,\overline{j})$$

$$= -1.5\,\overline{j} + 0.6\,\overline{i},$$

$$\text{or } \overline{v} = 0.6\,\overline{i} - 1.5\,\overline{j}$$

$$\text{Now } \overline{a_n} = \overline{\omega} \times (\overline{\omega} \times \overline{r}) = \overline{\omega} \times \overline{v}$$

$$\text{or } \overline{a_n} = -3\,\overline{k} \times (0.6\,\overline{i} - 1.5\,\overline{j})$$

$$= -1.8\,\overline{j} - 4.5\,\overline{i},$$

$$\text{or } \overline{a_n} = -4.5\,\overline{i} - 1.8\,\overline{j}\ m/s^2$$

$$\text{Now } \overline{a_t} = \overline{\alpha} \times \overline{r}$$

$$= 5k \times (0.5\,\overline{i} + 0.2\,\overline{j})$$

$$= 2.5\,\overline{j} - 1\,\overline{i},$$

$$\text{or } \overline{a_t} = -1\,\overline{i} + 2.5\,\overline{j}\ m/s^2$$

$$\overline{a} = \overline{a_n} + \overline{a_t} = -5.5\,\overline{i} + 0.7\,\overline{j}\ m/s$$

$$\text{Hence, } v = \sqrt{(0.6^2 + 1.5^2)} = 1.616\ m/s$$

$$\text{and } a = \sqrt{(5.5^2 + 0.7^2)} = 5.544\ m/s^2$$

Problems 6.1

1. A car wheel is rotating in a clockwise direction with an initial speed of 100 rev/min, when suddenly it decelerates with a constant magnitude of deceleration. If the car wheel takes 50 revolutions to stop, determine:

   (i) the magnitude of its deceleration;
   (ii) the time taken for the wheel to stop.

   [-0.1745 rad/s^2, 1 minute]

2. Two disks, namely disk 'A' and disk 'B', rotate together, but in opposite directions without slipping, as shown in Figure 6.11.

   If during a test, disk 'A' starts from rest with an angular acceleration of 1.5 rad/s^2, determine the time it takes for disk 'B' to attain an angular velocity of 30 rad/s.

   [6.667s]

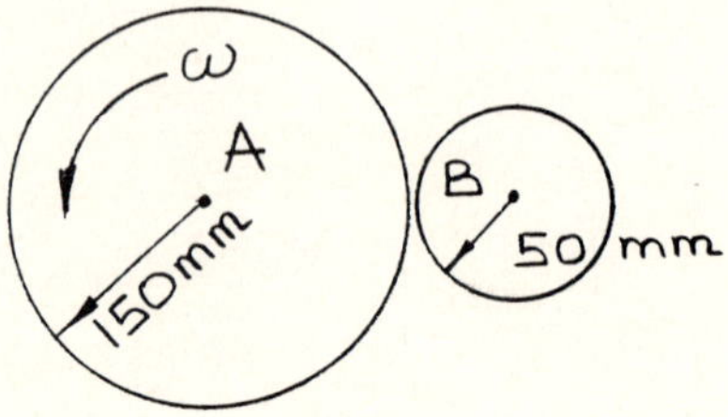

Figure 6.11

3. One end of a long light cord is attached to a weight, while the other end is wrapped around a drum. If the weight accelerates in a downward direction with a constant acceleration of 5 m/s^2, determine the angular acceleration and angular velocity of the drum after 6 seconds, assuming the system was initially at rest.

   [6.25 rad/s^2, 21.71 rad/s]

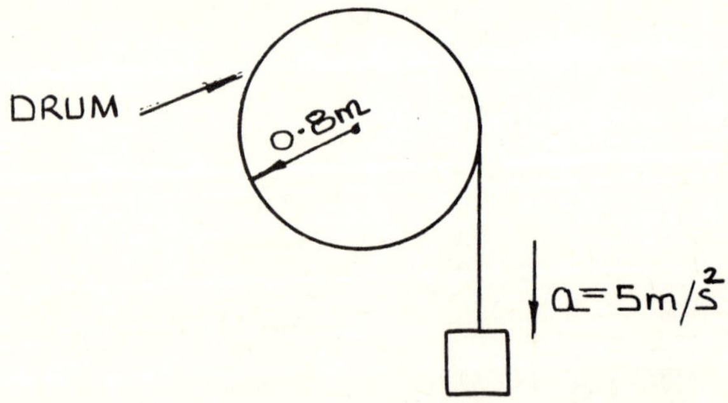

Figure 6.12

4. A wheel rotates, so that the counter-clockwise angular position of a straight line on it is given by the equation: $\theta = 4t^3 + t^2 + 5$ rad, where 't' is measured in seconds. Determine the angular displacement, when the angular acceleration increases from 28 to 56 rad/s^2.

   [44.37 rads]

5. A rectangular plate rotates about a fixed point 'O' in a clockwise direction, as shown in Figure 6.13. If the side BC has a constant angular velocity of 5 rad/s, determine the velocity and acceleration vectors of the point 'A'.

[$0.625\bar{i} - 1\,\bar{j}$ m/s, $-5\bar{i} - 3.125\bar{j}$ m/s^2]

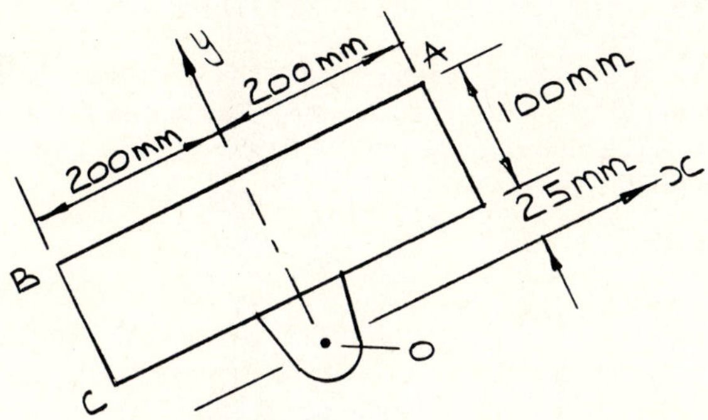

Figure 6.13

6. A plate in the form of an equilateral triangle rotates about the point 'O' in a counter-clockwise direction, as shown in Figure 6.14. If the normal and tangential acceleration of the plate centroid are 60 and 20 m/s^2, determine the angular velocity and angular acceleration of the plate.

[32.24 rad/s, 346.4 rad/s^2]

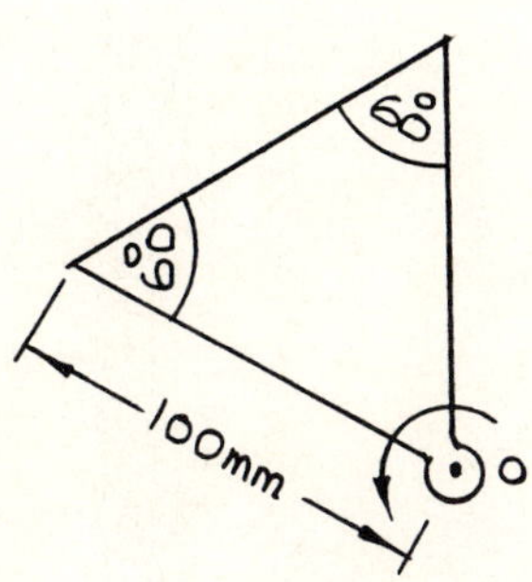

Figure 6.14

6.5 ABSOLUTE GENERAL PLANE MOTION ANALYSIS

In this section, we consider general plane motion in absolute terms. In general plane motion, both rotation and translation take place, and in many cases these two motions can be related to a fixed point, that is, absolute motion. Firstly, the geometry of the motion has to be obtained, and after this, the velocities and accelerations can be determined through differentiation. Very often, it is convenient to link translatory motion to angular motion, that is, $s = f(\theta)$.

Sometimes, however, the motion is too complex to be described in absolute terms and for these cases, it is usually easier to relate the motion in relative terms.

In absolute terms, it is important that the mathematical derivations are consistent. For example, if the rotation 'θ' is said to be positive in a counter-clockwise direction the angular velocity and angular accelerations, namely $\dot{\theta}$ and $\ddot{\theta}$, respectively, must be said to be positive in a counter-clockwise direction as well.

Example 6.4

A wheel rotates on a flat horizontal surface in the 'x' direction, without slipping, as shown in Figure 6.15, so that its centre 'O' moves to the position 'O^1'. Determine an expression for the translatory motion of 'O' in terms of the rotary motion. Hence, or otherwise, determine expressions for the velocity and acceleration of the rim of the wheel, as the rim comes in contact with the horizontal surface.

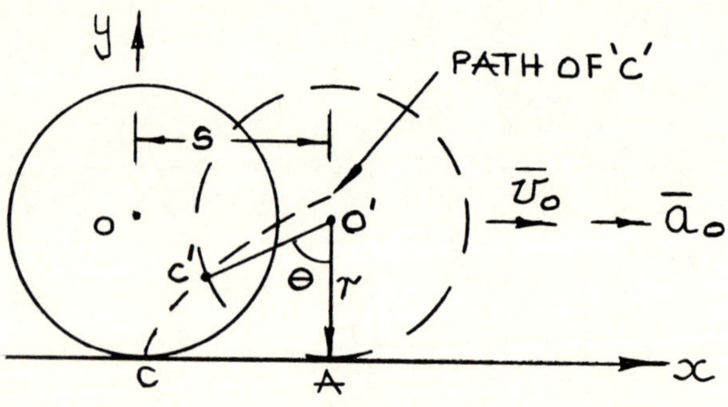

Figure 6.15

Figure 6.15 shows the motion of a wheel, including the locus of the motion of 'C'. As the wheel does not slip, the distance 's' is equal to the arc length AC^1, that is

$$s = r\theta \tag{6.18}$$

$$\text{Now } \overline{v_o} = \dot{s} = r\dot{\theta} = r\omega \tag{6.19}$$

$$\text{and } \overline{a_o} = \ddot{s} = r\ddot{\theta} = r\alpha \tag{6.20}$$

Motion of Rim

Consider the path of point C along to C^1,

$$x = \text{horizontal distance from C to } C^1$$
$$= s - r\sin\theta$$
$$= r\theta - r\sin\theta \qquad (6.21)$$
$$\text{or } x = r(\theta - \sin\theta)$$

Similarly

$$y = \text{vertical distance from C to } C^1$$
$$= r - r\cos\theta \qquad (6.22)$$
$$= r(1 - \cos\theta)$$

Differentiating equation (6.21), with respect to '*t*', we get

$$\dot{x} = r\left(\dot{\theta} - \dot{\theta}\cos\theta\right)$$
$$= r\,\dot{\theta}\,(1 - \cos\theta),$$
$$\text{or } \dot{x} = \overline{v_o}\,(1 - \cos\theta) \qquad (6.23)$$
$$\text{and } \ddot{x} = \overline{\dot{v}_o}\,(1 - \cos\theta) + \overline{v_o}\,\dot{\theta}\sin\theta$$
$$= \overline{a_o}\,(1 - \cos\theta) + r\,\omega^2\sin\theta$$

Differentiating equation (6.22) with respect to 't' we get

$$\dot{y} = r\left(\dot{\theta}\sin\theta\right)$$
$$= \overline{v_o}\sin\theta \qquad (6.24)$$

$$\text{and } \ddot{y} = \overline{\dot{v}_o}\sin\theta + \overline{v_o}\,\dot{\theta}\cos\theta$$
$$= \overline{a_o}\sin\theta + r\,\omega^2\cos\theta \qquad (6.25)$$

It should be noted that when $\theta = o$,

$$\dot{x} = o \quad \text{and} \quad \dot{y} = o$$

$$\text{Also } \ddot{x} = o \quad \text{and} \quad \ddot{y} = r\,\omega^2$$

That is, because the velocity at the point 'C' is zero, this point is called the **instantaneous**

centre of zero velocity. This point is particularly useful for determining other velocities within the wheel, as momentarily, the body rotates about the instantaneous centre of zero velocity.

Problems 6.2

1. A cotton reel rolls to the right on the flat horizontal surface, when it is pulled by a thread. If the velocity of the moving thread is 0.2 m/s, determine the linear and angular velocity of the reel.

   [0.333 m/s, 26.67 rad/s]

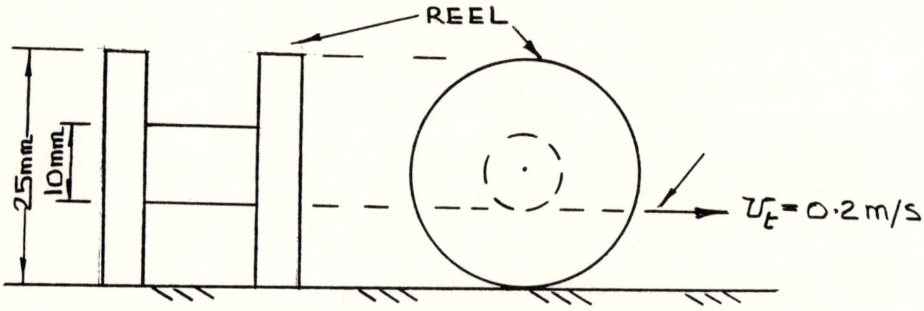

(a) end elevation **(b) view from side**

Figure 6.16

2. The wheels of a diesel electric locomotive are powered through the gearing system shown in Figure 6.17. If the speed of the locomotive is 80 km/h, determine the angular velocity of the power gears 'A', where the gear 'B' is an integral part of the wheel. Neglect slipping.

   [74.08 rad/s]

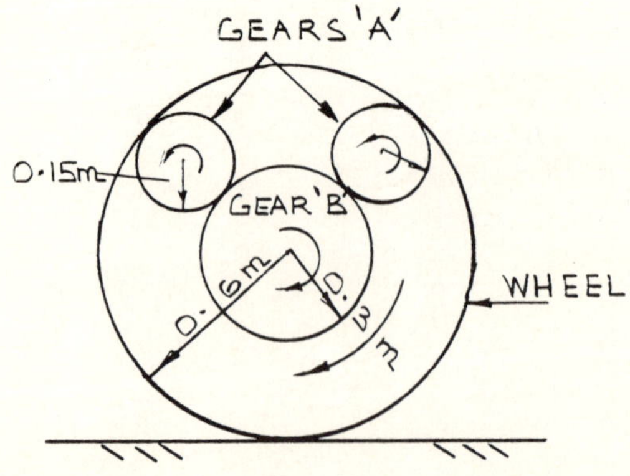

Figure 6.17

3. The cam shown in Figure 6.18 rotates about the point 'O' and at the same time, the cam follower keeps in contact with the cam throughout its motion. Determine expressions for the velocity and acceleration of the cam follower at the point 'C' at any angle 'θ'.

[$v = -2\,r\,\omega\,.\,\sin\theta$ m/s, $a = -2r\,(\omega^2 \cos\theta + \alpha \sin\theta)$]

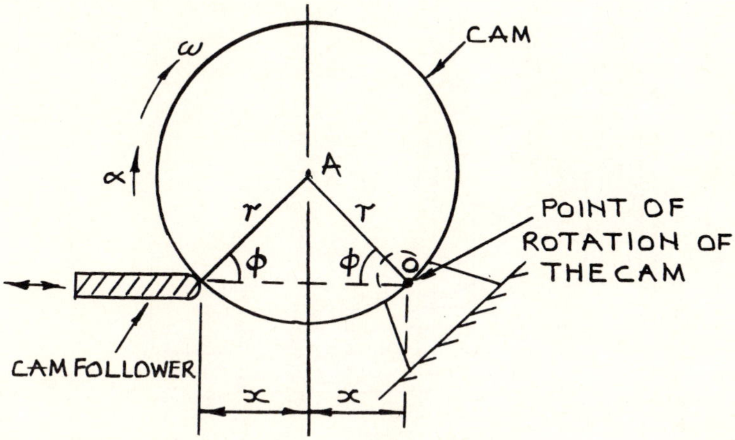

Figure 6.18

4. The device shown in Figure 6.19 is used for converting rotary motion of the crank OA into reciprocating motion of the rod BC. If the crank OA of length 'r' is rotating about the fixed point 'O', with a constant velocity 'ω^1', what will be the velocity and acceleration of the rod BC.

[$v_{BC} = \mathrm{r}\,\omega\,\cos\theta$ m/s, $a_{BC} = -\mathrm{r}\,\omega^2 \sin\theta$ m/s^2]

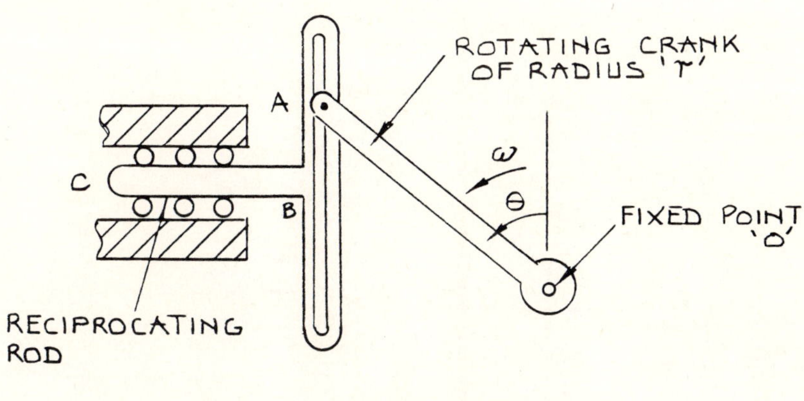

Figure 6.19

5. The artefact of Figure 6.20 shows that the top end of the bar AB is restricted to vertical motion in a slider and the bottom end of the bar is restricted to horizontal motion in another slider. If the horizontal velocity at the bottom of the bar is 5 m/s, determine the angular velocity of the bar and the vertical velocity of the top end of the bar when $\theta = 30°$.

[-6.667 rad/s, -8.66 m/s]

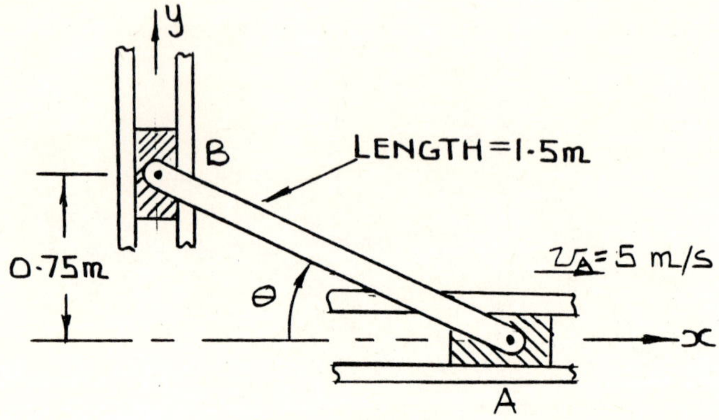

Figure 6.20

6. A crank-slider mechanism is shown in Figure 6.21, where the crank OA rotates about the fixed point 'O', with an angular velocity of ω_o. Determine expressions for the angular velocity and acceleration of the conrod AB. Assume that the angular velocity and acceleration of the conrod are positive in the counter-clockwise direction.

[$\omega_{AB} = r\,\omega_o \cos\theta \,/\, \{\ell[(1 - r \sin\theta \,/\, \ell)^2]^{1/2}\}$,

$\alpha_{AB} = r\,\omega_o^2 \sin\theta\, [(r/\ell)^2 - 1] \,/\, \{\ell[1 - (r \sin\theta \,/\, \ell)^2]^{3/2}$

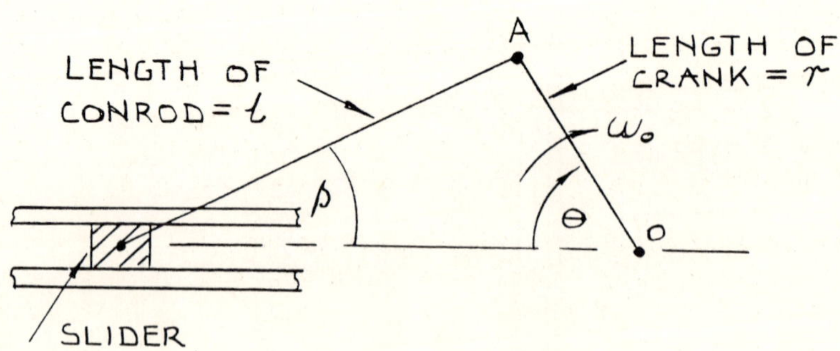

Figure 6.21

6.6 RELATIVE MOTION ANALYSIS: VELOCITY

In Section 6.3, it was shown that for certain problems in general plane motion, absolute motion analysis was suitable. This was especially so when the geometry of the position vectors was simple. However, for other cases, especially for when the geometry of the position vectors is not straight forward, it is often more suitable to use relative motion analysis.

In this case, two systems of axes are used, one is fixed and the other appears on the rigid body, as shown in Figure 6.22.

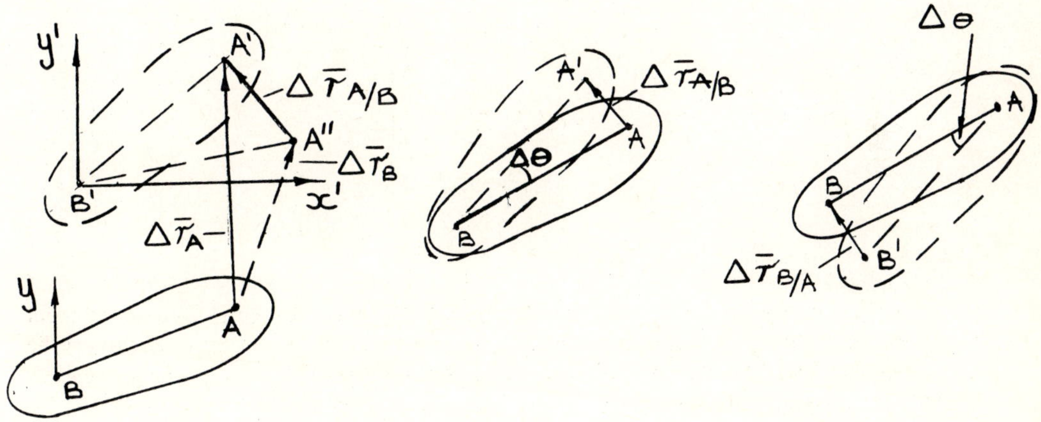

(a) **(b)** **(c)**

Figure 6.22

Consider the straight line AB on the rigid body shown in Figure 6.22(a), which has a fixed reference axes going through the point 'B'. Now the rigid body translates and rotates to the position shown by the dashed figure of Figure 6.22(a) in the time 'Δt'. In the dashed figure of Figure 6.22(a), the new position of the line AB is A^1 B^1, where the moving axes x^1 - y^1 has its origin at B^1. It can be seen that the line A^1 B^1 rotates with respect to the line B^1 A^{11}; this effect is also shown in Figure 6.22(b).

Now if translation is separated from rotation the line AB moves to the line A^{11} B^1, due to translation, as shown in Figure 6.22(a). It is evident therefore, that if 'B' is made the reference point, as shown in Figure 6.22,

$$\Delta\overline{r_A} = \Delta\overline{r_B} + \Delta\overline{r_{A/B}} \qquad (6.26)$$

where

$\Delta\overline{r_A}$ = translational displacement of `A'

$\Delta\overline{r_B}$ = translational displacement of `B'

$\Delta\overline{r_{A/B}}$ = rotational displacement of `A' relative to B = r.Δθ,

as shown in Figure 6.22(a) and (b).

Alternatively if 'A' is fixed and 'B' is drawn relative to 'B', as shown in Figure 6.22(c), then

$$\Delta\overline{r_{B/A}} = -\Delta\overline{r_{A/B}}$$

where

$\Delta\overline{r_{B/A}}$ = rotational displacement of B relative to A

Now as the motion takes place in a time Δt, we have from equation (6.26)

$$\frac{\Delta\overline{r_A}}{\Delta t} = \frac{\Delta\overline{r_B}}{\Delta t} + \frac{\Delta\overline{r_{A/B}}}{\Delta t}$$

or

$$\overline{v_A} = \overline{v_B} + \overline{v_{A/B}}$$

where

$\overline{v_A}$ = velocity of `A'

$\overline{v_B}$ = velocity of `B'

$\overline{v_{A/B}}$ = rotational relative velocity of `A' with respect to the fixed axis `B'

$= r\,.\,\omega$

In vector quantities

$$\overline{v_{A/B}} = \overline{\omega} \times \overline{r} \tag{6.28}$$

Equation (6.22) is shown with the aid of Figure 6.23.

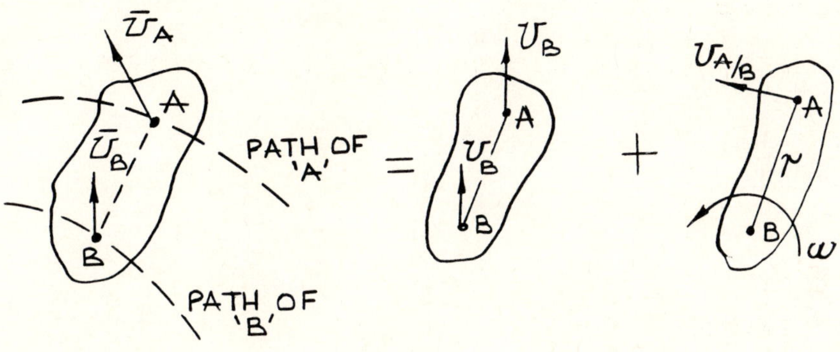

(a) (b) (c)

Figure 6.23

The velocity diagram is shown in Figure 6.24.

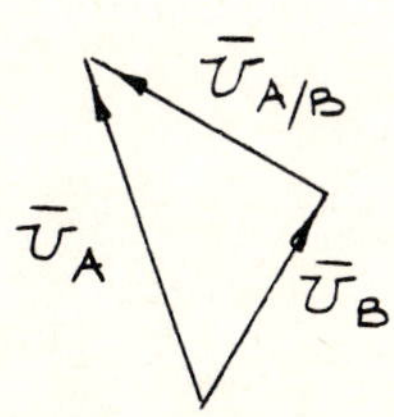

Figure 6.24 Velocity Diagram

Example 6.5

A wheel of radius 400 mm rolls on a flat horizontal surface to the right of Figure 6.25, without slipping. If the velocity of the centre of gravity of the wheel is v_o = 2.5 m/s, determine the velocity of a point 'A', which is on a radius of 300 mm.

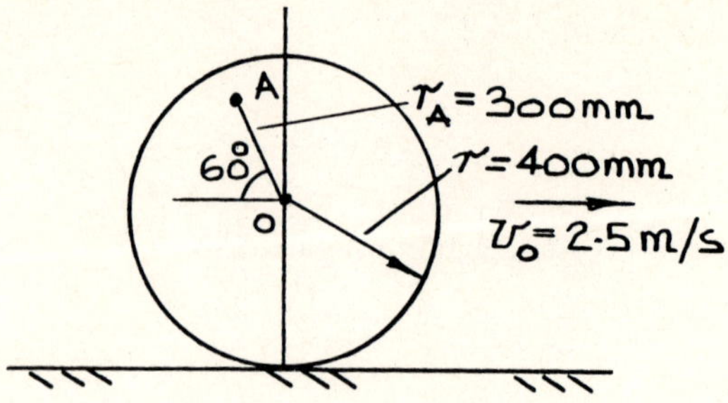

Figure 6.25

Now

$$v_A = v_o + v_{A/o}$$

where

$$v_{A/o} = \omega \times r_A$$

Now $v_o = \omega r$

or $2.5 = \omega \times 400 \times 10^{-3}$

or $\omega = 6.25$ rad/s

Hence, $v_{A/o} = 6.25 \times 300 \times 10^{-3}$

$= 1.875$ m/s, which is perpendicular to OA

The velocity diagram is shown in Figure 6.26.

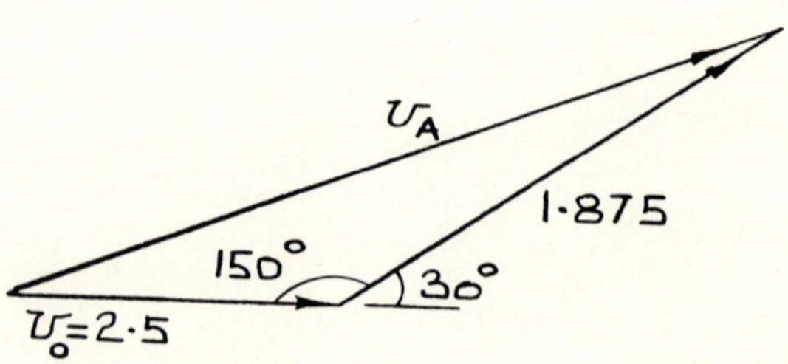

Figure 6.26

From the cosine formula of equation (1.7)

$$v_A^2 = 2.5^2 + 1.875^2 + 2 \times 2.5 \times 1.875 \cos 150°$$

$$= 6.25 + 3.516 + 8.119$$

$$\text{or } v_A = 4.229 \; m/s$$

Vector Analysis Approach

Another solution to this problem is through vector analysis, as follows:

$$\overline{v_A} = \overline{v_o} + \overline{v_{A/o}}$$
$$= \overline{v_o} + \overline{\omega} \times \overline{r_A} \quad (6.29)$$

$$\text{Now } \overline{\omega} = -6.25k \quad (6.30)$$

It should be noted that the sign of ω is negative because it is clockwise looking downwards into the plane of the paper.

$$\overline{r_A} = 0.3 \left(-\bar{i} \cos 60° + \bar{j} \sin 60°\right)$$

$$= -0.15 \; \bar{i} + 0.26 \; \bar{j}$$

$$\text{Now } \overline{v_o} = 2.5\bar{i} \quad (6.31)$$

Substituting equation (6.30) and (6.31) into (6.29), we get

$$\overline{v_A} = 2.5 \; \bar{i} - 6.25\bar{k} \times \left(-0.15 \; \bar{i} + 0.26\bar{j}\right)$$

$$= 2.5\bar{i} + 0.938\bar{j} + 1.625\bar{i}$$

$$\text{or } \overline{v_A} = 4.125\bar{i} + 0.938\bar{j} \quad (6.32)$$

$$\text{or } v_A = \sqrt{(4.125^2 + 0.938^2)}$$

$$= 4.23 \text{ m/s (as before)}$$

Example 6.6

A link is restricted to the motion shown in Figure 6.27, where the top end moves in a vertical slider and the lower end moves in a horizontal slider. If the downward velocity of 'A' is 3 m/s, what is the velocity of 'B', when $\theta = 60°$?

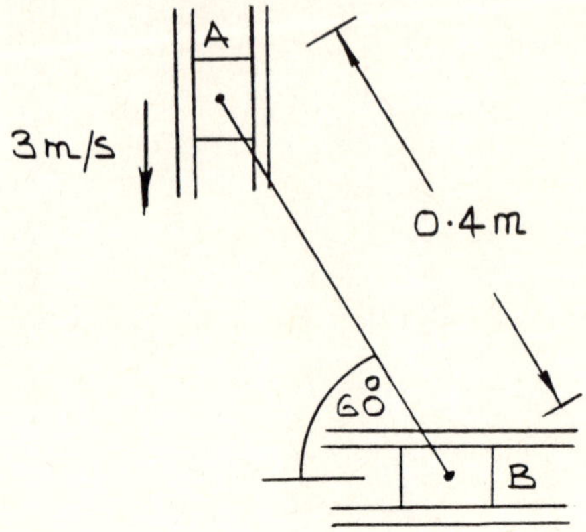

Figure 6.27

$$\overline{v_A} = \overline{v_B} + \overline{v_{A/B}}$$

$$= \overline{v_B} + \overline{\omega} \; x \; \overline{r_{A/B}}$$

$$\text{or } -3\bar{j} = v_B \, \bar{i} + \omega\bar{k} \; x \; (-0.4 \, \bar{i} \cos 60 + 0.4\bar{j} \sin 60)$$

$$= v_B \, \bar{i} + \omega\bar{k} \; x \; (-0.2 \, \bar{i} + 0.346\bar{j})$$

$$-3\bar{j} = v_B \, \bar{i} - 0.2 \, \omega\bar{j} - 0.346 \, \omega\bar{i}$$

Equating $\bar{j}$'s

$$\omega = \frac{3}{0.2} = 15\text{rad/s. (Counter-clockwise)}$$

Equating i's

$$O = v_B - 0.346 \; x \; 15$$

$$\therefore \underline{v_B = 5.19 \text{ m/s}}$$

Example 6.7

A three bar linkage is shown in Figure 6.28, where the points 'A' and 'D' are firmly pinned, so that the linkage can rotate about these fixed axes. The linkage members are also pinned together at the points 'B' and 'C'.

If the bar AB is subjected to a clockwise angular velocity of 20 rad/s when $\theta = 70°$, determine the angular velocities of links BC and CD at this instant.

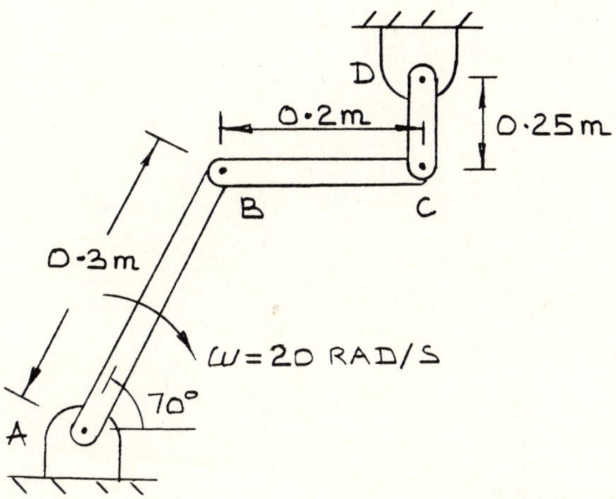

Figure 6.28

$$\overline{v_B} = \overline{\omega_{B/A}} \ x \ \overline{r_B}$$

The angular velocity of AB is negative because it is clockwise looking into the plane of the paper.

$$\therefore \overline{v_B} = -20\bar{k} \ x \ (0.3 \ \bar{i} \ . \ \cos 70° + 0.3 \ \bar{j} \ . \ \sin 70)$$

$$= -2.052 \ \bar{j} + 5.638 \ \bar{i}$$

$$\text{or } \overline{v_B} = (5.638 \ \bar{i} - 2.052 \ \bar{j})$$

Now

$$\overline{v_c} = \overline{v_B} + \overline{\omega_{B/C}} \times \overline{r_{C/B}}$$

$$\text{or } v_C \cdot \bar{i} = (5.638\,\bar{i} - 2.052\,\bar{j})$$

$$+ \omega_{B/C} \cdot \bar{k} \times (0.2i + 0)$$

or

$$v_C\,\bar{i} = 5.638\,\bar{i} - 2.052\,\bar{j} + 0.2\,\omega_{B/C}\,\bar{j}$$

$$= 5.638\,\bar{i} + (0.2\,\omega_{B/C} - 2.052)\,\bar{j}$$

Equating i's, we get

$\underline{v_c = 5.638 \text{ m/s}}$

Equating j's, we get

$o = 0.2\,\omega_{B/C} - 2.052$

or $\underline{\omega_{B/C} = 10.26 \text{ rad/s}}$ (Counter-clockwise)

As links CD rotates about D

$$\overline{v_C} = \overline{\omega_{C/D}} \times \overline{r_C}$$

$$\text{or } 5.638\,\bar{i} = \omega_{C/D}\bar{k} \times (-0.15\bar{j})$$

$$= 0.15\omega_{C/D} \cdot \bar{i}$$

$$\text{or } \underline{\omega_{C/D}} = \underline{37.39 \text{ rad/s } (\textit{Counter-clockwise})}$$

Problems 6.3

1. A rocket is launched vertically upwards but due to a malfunction of one of its engines, it rotates in a counter-clockwise direction with an angular velocity of 3.5 rad/s. If at the same time, the upward velocity of its centre of gravity is 70 m/s, what is the magnitude and direction of the velocity of its nose.

   [v = 75.39 m/s @ 68.2° to the horizontal]

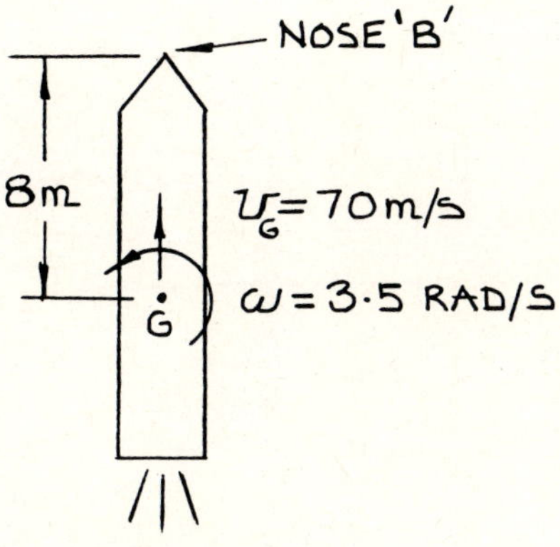

Figure 6.29

2. A link is constrained to the motion shown in Figure 6.30, where at a certain time, the upper end 'B' moves in a horizontal direction with a velocity of 1.5 m/s, while the lower end of the link moves in a vertical guide. Determine the velocity of A and the angular velocity of the link.

[v_A = -0.75 m/s; ω = 0.75 rad/s (Counter-clockwise)]

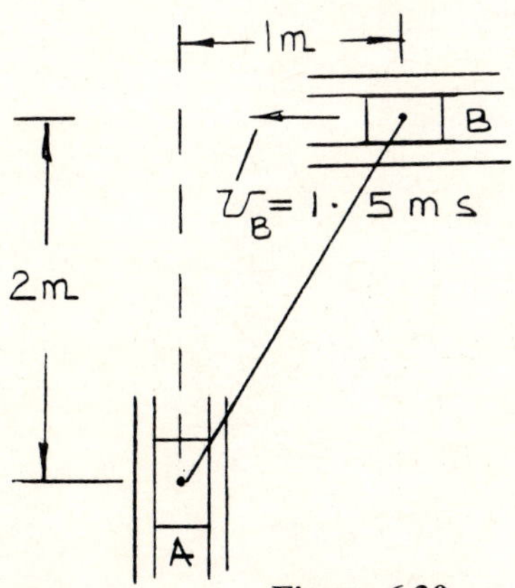

Figure 6.30

3. A telescopic link AB is rotated by two rotating driving links AC and DB. Determine the angular velocity of the link AB.

[0.373 rad/s]

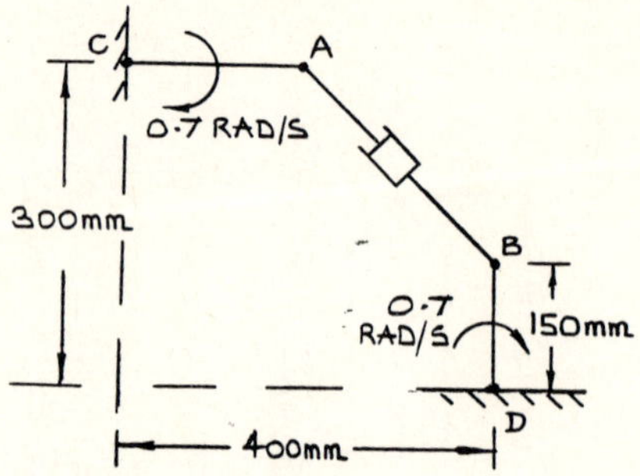

Figure 6.31

4. The linkage system shown in Figure 6.32 is subjected to a horizontal velocity of 0.7 m/s at the point 'C'. If the link AB rotates about the fixed axis 'A', determine the angular velocities of the link AB and BC, when these two links are perpendicular to each other.

[ω_{AB} = 6.5 rad/s; ω_{BC} = 2.6 rad/s (both Counter-clockwise)]

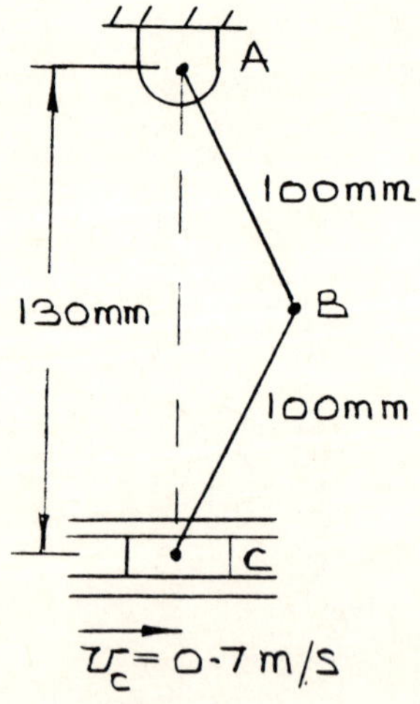

Figure 6.32

5. The mechanism shown in Figure 6.33, consists of two links AB and BC, where link AB rotates with an angular velocity of 5 rad/s about the fixed axis 'A'. The link BC is pinned to the link AB at the point 'B' and the link BC is constrained to move horizontally at the point 'C'. Determine the velocity at 'C'.

[866 mm/s]

6. A four bar linkage is shown in Figure 6.34, where 'A' and 'D' are fixed axes of rotation. If link AB is rotating with a counter-clockwise angular velocity of 7 rad/s, determine the angular velocities of links BC and CD, when the link CD lies vertically with the 'x' axis.

[ω_{BC} = 1.2 rads; ω_{CD} = 3.947 rad/s (both Counter-clockwise)]

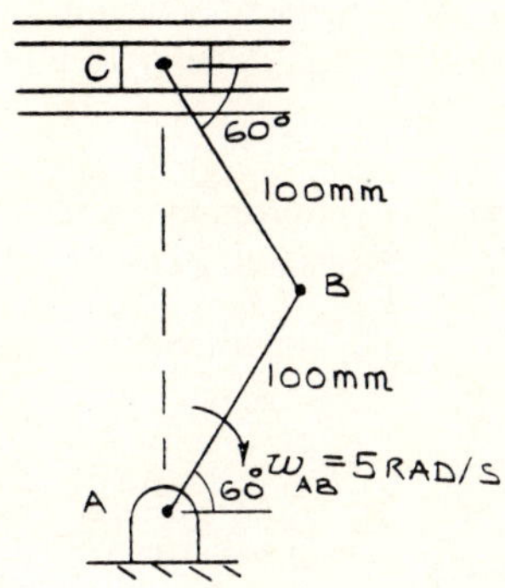

Figure 6.33

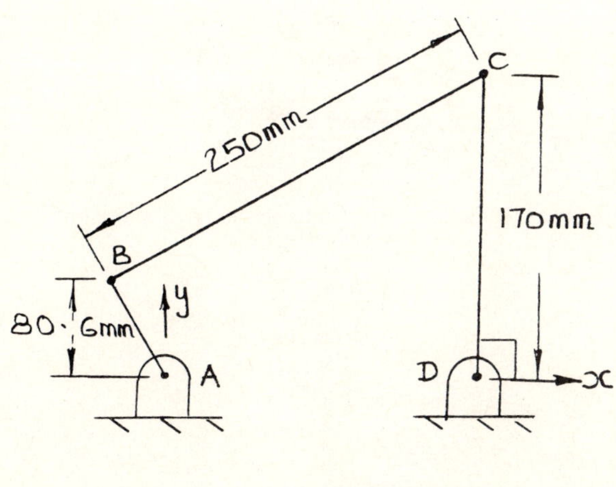

Figure 6.34

6.7 INSTANTANEOUS CENTRE OF ZERO VELOCITY

In the general plane motion of a rigid body, it can be shown that at any instant in time, the body has an **instantaneous centre of zero** velocity, where the body rotates momentarily, about the point. The instantaneous centre of zero velocity need not be on the body itself.

To demonstrate the position of the instantaneous centre of zero velocity, consider the motion of the rigid body shown in Figure 6.35(a).

In Figure 6.35(a), the points 'A' and 'B' move momentarily in circular paths, which are not parallel to each other. The directions of the velocities of 'A' and 'B', namely, v_A and v_B are drawn on the Figure. Now if at 'A' and 'B', perpendicular lines are drawn to the directions of the velocity vectors v_A and v_B, these perpendicular lines intersect at the point 'C', which is the instantaneous centre of zero velocity. The rigid body is said to momentarily rotate about 'C'.

Similarly, in the case where the point 'A' moves in a direction parallel to the motion of 'B', as shown in Figures 6.35(b) and (c), the instantaneous centre of zero velocity is given by the points 'C'.

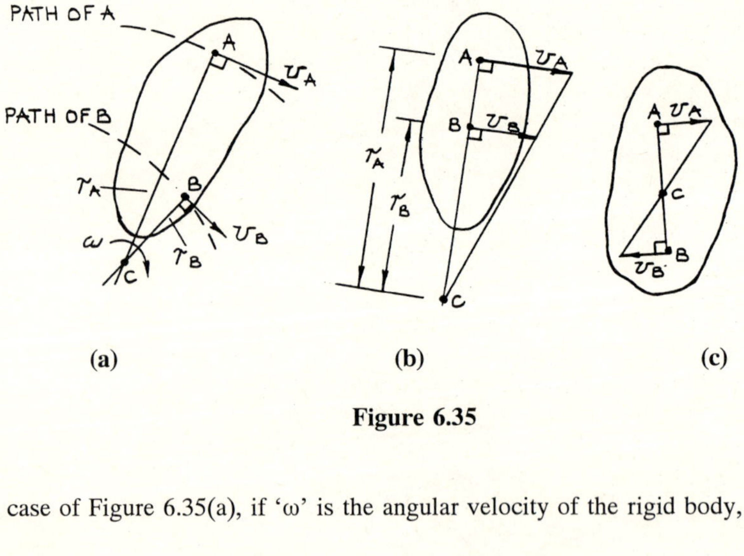

(a) (b) (c)

Figure 6.35

In the case of Figure 6.35(a), if 'ω' is the angular velocity of the rigid body, then,

$$v_A = \omega r_A \tag{6.34}$$

and $$v_B = \omega r_B \tag{6.35}$$

In the cases of Figures 6.35(b) and (c), if 'ω' is known v_A and v_B can be obtained from the above equations, together with the properties of similar triangles.

Example 6.8

A wheel rotates on a flat horizontal surface, to the right of the figure, without slipping, as shown in Figure 6.36. Determine the position of the instantaneous centre of zero velocity and the velocity of the point 'A' on the wheel.

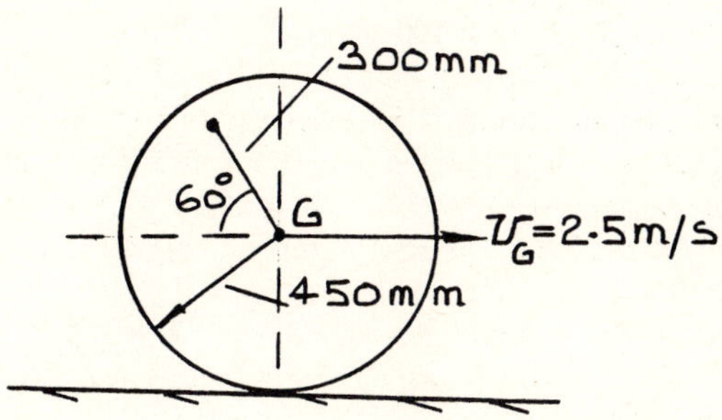

Figure 6.36

By inspection, it can be seen that the instantaneous centre of zero velocity is at the position 'C'. The reason for this is that the centre of gravity is travelling to the right with a velocity of 2.5 m/s. However, due to rotation above, the velocity component is 2.5 m/s to the left, making the net velocity equal to zero at 'C'.

Knowing the position 'C', the velocity of 'A' can be obtained with the aid of Figure 6.37.

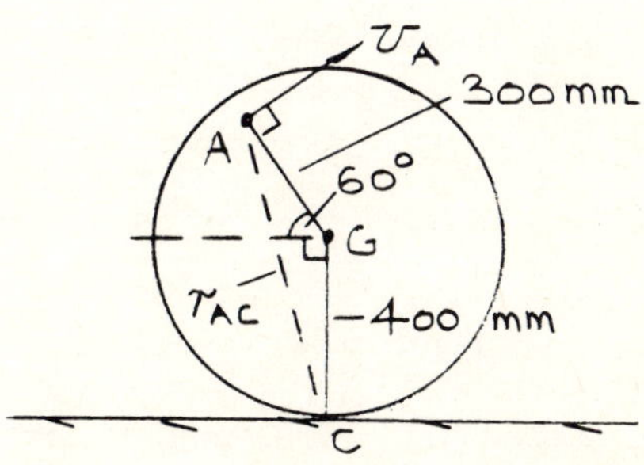

Figure 6.37

$$\text{Now } \omega = \frac{2.5}{400 \times 10^{-3}} = 6.25 \text{ rad/s} \tag{6.36}$$

$$\text{and } v_A = \omega \times r_{AC}$$

From the cosine formula of equation (1.7)

$$r_{AC}^2 = 300^2 + 400^2 - 2 \times 300 \times 400 \cos(60° + 90°)$$

$$= 250{,}000 - 240{,}000 \cos 150°$$

$$\text{or } \underline{r_{AC}} = \underline{676.6 \text{ mm}} \tag{6.37}$$

Substituting the values for ω and r_{AC} into equation (6.36), we get

$$v_A = 6..25 \times 676.6 \text{ mm/s}$$

$$= 4.229 \text{ m/s}$$

Example 6.9

A four bar chain is shown in Figure 6.38, where the link CD rotates with an angular velocity of 7 rad/s.

Determine the velocity at B and E and the angular velocity of the link BF.

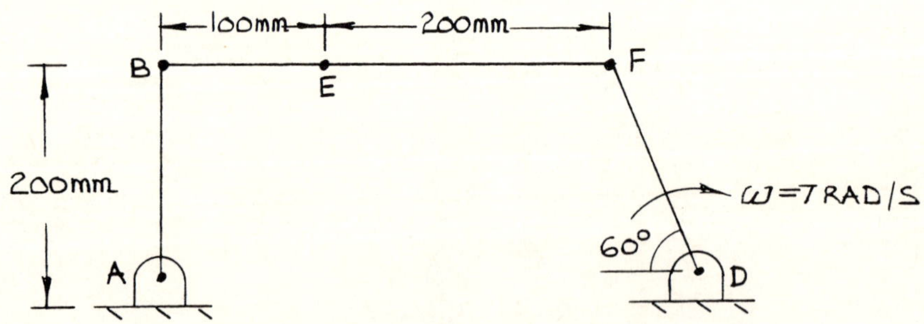

Figure 6.38

Length of FD = ℓ_{FD} = 200/sin 60° = 230.9 mm

As FD rotates about a fixed point, namely 'D', the velocity at 'F', namely v_F, is perpendicular to FD.

$$v_F = \omega \times \ell_{FD} = 7 \times 230.9 = 1616.6 \text{ mm/s}$$

Similarly it can be seen from Figure 6.38, that the velocity at 'B', namely v_B, is perpendicular to the bar AB. Hence, the position of the instantaneous centre, namely 'C' can be determined with the aid of Figure 6.39.

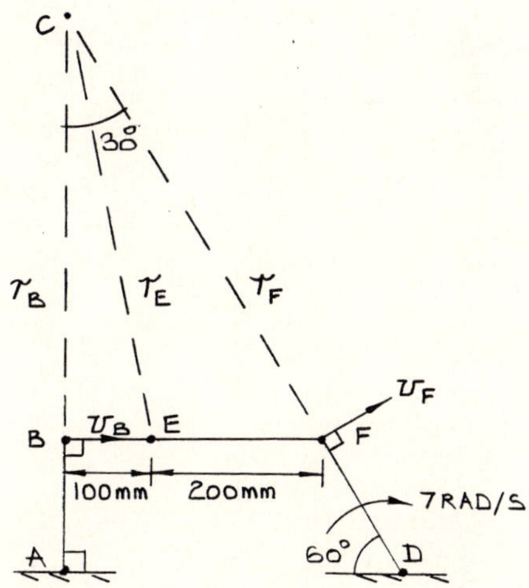

Figure 6.39

$$\frac{BC}{BF} = \tan 60°$$

$$\therefore r_B = BC = 300 \tan 60°,$$

$$\text{or } \underline{r_B} = \underline{519.6 \text{ mm}}$$

Similarly

$$r_C^2 = r_B^2 + BF^2$$

$$= 519.6^2 + 300^2$$

$$\text{or } r_C = 600 \text{ mm}$$

Now, ω_{BF} = angular velocity of BF

$$= \frac{v_F}{r_F}$$

$$= \frac{1616.6}{600}$$

or ω_{BF} = 2.694 rad/s

Now, $v_B = r_B \ x \ \omega_{BF}$

$= 519.6 \ x \ 2.694$

or v_B = 1400 mm/s

To calculate v_E

From Figure 6.39

$$r_E^2 = r_B^2 + BE^2$$

$$= 519.6^2 + 100^2$$

or r_E = 529.1 mm

Now $v_E = r_E \ x \ w_{BF}$

$= 529.1 \ x \ 2.694$

or v_E = 1425.5 mm/s which is perpendicular to CE

Problems 6.4

1. A slider-crank mechanism is shown in Figure 6.40, where the slider 'D' moves to the right of the figure with a velocity of 2.5 m/s. If the crank AB rotates about the fixed axis 'A', determine the angular velocities of the crank and the conrod BD.

   [ω_{AB} = 2.166 rad/s (cw); ω_{BD} = 0.722 rad/s (ccw)]

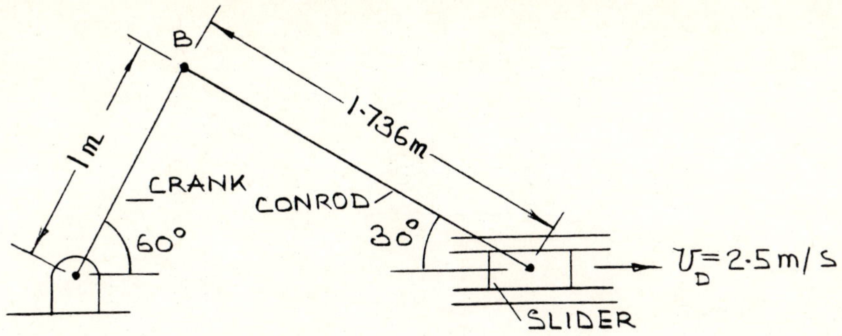

Figure 6.40

2. A cylinder of diameter 1 m rolls between two moving plates, without slipping, as shown in Figure 6.41. If the plate on the top of the cylinder is moving to the right with a velocity of 0.5 m/s and the plate on the bottom of the cylinder moves to the left with a velocity of 0.25 m/s, determine the angular velocity of the cylinder and the velocity of its centre of gravity.

[ω = 0.75 rad/s (cw); v_G = 0.125 m/s (to the right)]

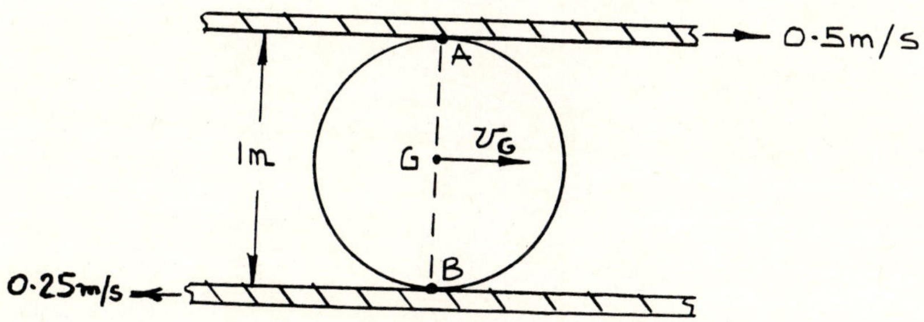

Figure 6.41

3. Solve problem No. 5 from Problems 6.2

[ω_{AB} = 6.667 rad/s (ccw); v_B = 8.667 m/s (downwards)]

4. A four bar chain is shown in Figure 6.42, where at a certain instance in time, the link DE rotates clockwise about the point 'E' with an angular velocity of 5 rad/s. Determine the angular velocities of the bars AB and BD.

[ω_{AB} = 4.99 rad/s (cw); ω_{BD} = 2.11 rad/s (ccw)]

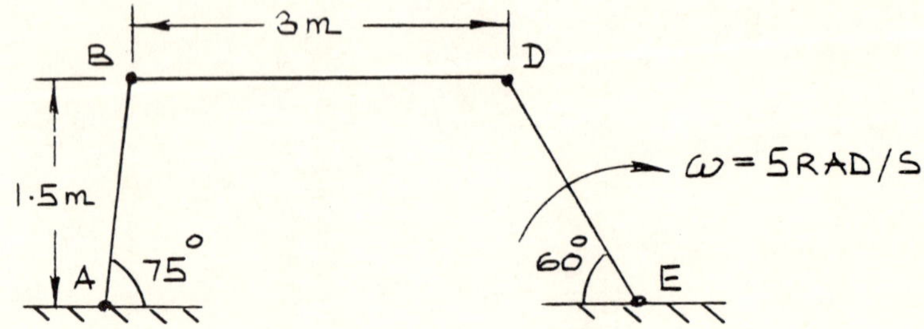

Figure 6.42

5. The linkage shown in Figure 6.43, rotates about the fixed axis 'A' & 'F'. If the points 'B' & 'E' are also pinned, and the link AB is rotating with an angular velocity of 2.5 rad/s, determine the velocity at 'E' and the angular velocity of the link BE.

[v_E = 4.194 m/s; ω_{BE} = 1.88 rad/s (ccw)]

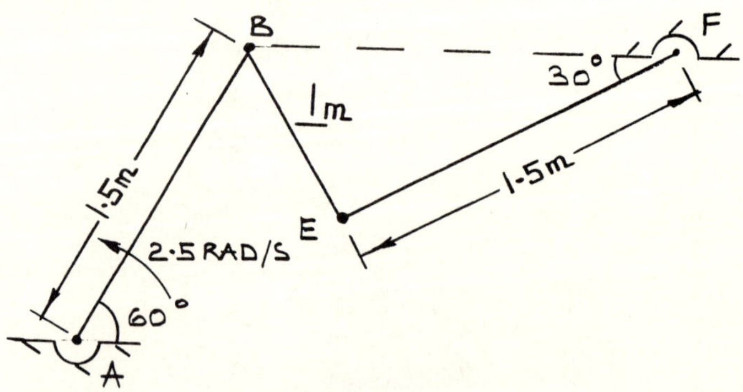

Figure 6.43

6. The piston in the linkage shown in Figure 6.44 subjects the link BE to a horizontal velocity, namely, v_D at the point 'D'. If the link AB rotates about the fixed point 'A', determine the velocity of the point 'E'.

$$\left[v_E = 0.5\, v_D \sqrt{\left\{\sec^2\left(\frac{\theta}{2}\right) + 8\right\}} \right]$$

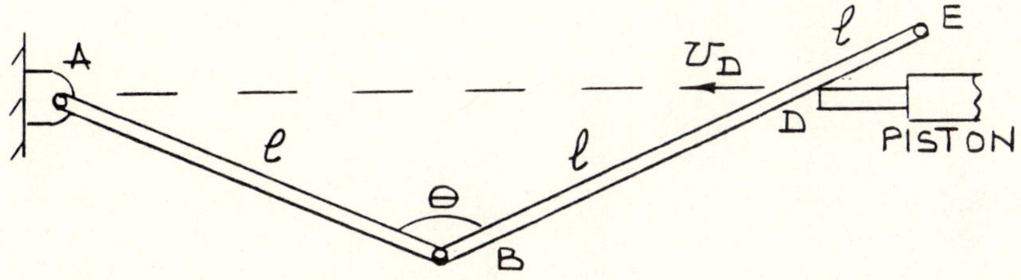

Figure 6.44

6.8 RELATIVE MOTION ANALYSIS: ACCELERATION

The relative acceleration equation can be explained by differentiating the relative velocity equation, as follows:

$$\overline{\dot{v}_A} = \overline{\dot{v}_B} + \overline{\dot{v}_{A/B}} \tag{6.38}$$

$$\text{or } \overline{a_A} = \overline{a_B} + \overline{a_{A/B}} \tag{6.39}$$

where

$\bar{a}_A$ = the acceleration of the point 'A', as shown in Figure 6.45

$\bar{a}_B$ = the acceleration of the point 'B'

$\bar{a}_{A/B}$ = the rotational acceleration of the point 'A' as seen from a non-rotating observer at 'B'.

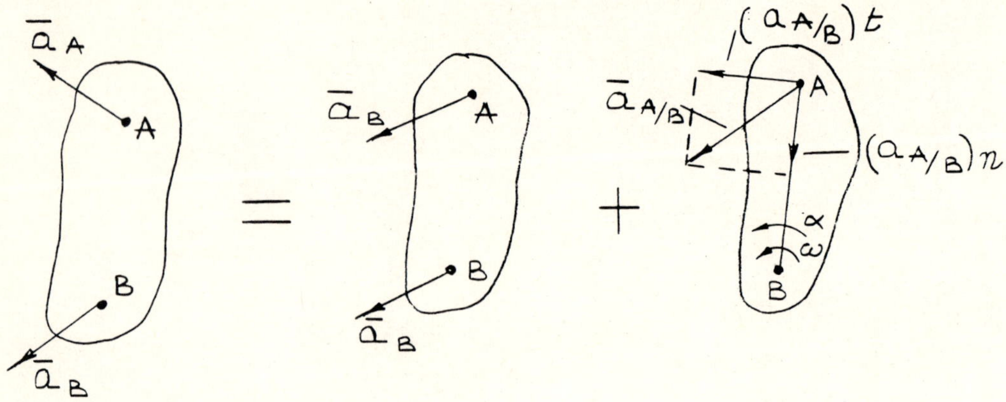

Figure 6.45

From Figure 6.45(c)

$$\overline{a_{A/B}} = \left(a_{A/B}\right)_n + \left(a_{A/B}\right)_t \tag{6.40}$$

where

$$\left(a_{A/B}\right)_n = \left(v_{A/B}\right)^2/r = r\omega^2 \tag{6.41}$$

$$\text{and } \left(a_{A/B}\right)_t = \dot{v}_{A/B} = \alpha r \tag{6.42}$$

'n' and 't' refer to the normal and tangential directions to AB

r = AB

ω = angular velocity

α = angular acceleration

In vector rotation, equations (6.41) and (6.42) appear as:

$$\left(\overline{a_{A/B}}\right)_n = \bar{\omega}\ x\ \left(\bar{\omega}\ x\ \bar{r}\right) \tag{6.43}$$

$$\left(\overline{a_{A/B}}\right)_t = \bar{\alpha}\ x\ \bar{r} \tag{6.44}$$

The vector diagram for these equations is shown in Figure 6.46.

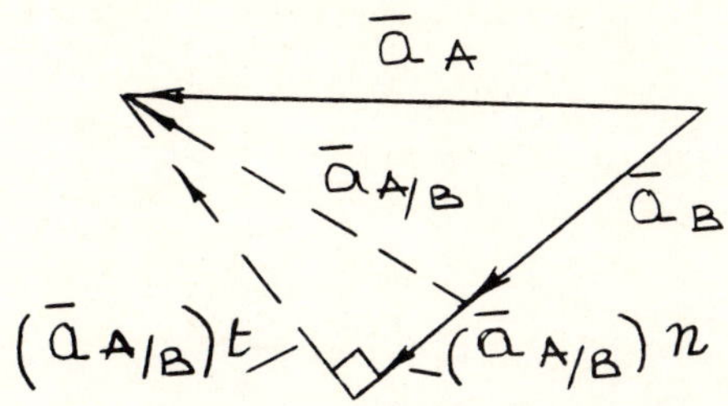

Figure 6.46 Vector Diagram

Example 6.10

The wheel shown in Figure 6.47 travels and accelerates to the right of the figure, without slipping. Determine expressions for the acceleration of the points 'A' and 'C' at the instant of the motion shown.

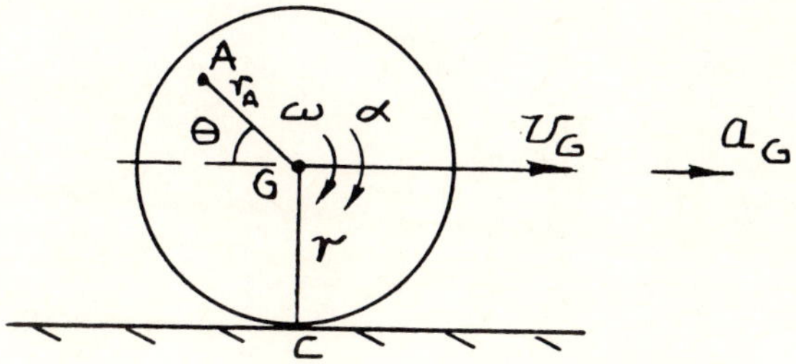

Figure 6.47

From equations (2.54) and (2.56)

$$\omega = v_G/r$$

and $$\alpha = a_G/r$$

From equation (6.39)

$$\overline{a_A} = \overline{a_G} + \overline{a_{A/G}}$$

$$= \overline{a_G} + (a_{A/G})_n + (a_{A/G})_t$$

where $\left(a_{A/G}\right)_n \;=\; r_A\,\omega^2 \;=\; r_A\left(\dfrac{v}{r}\right)$

and $\left(a_{A/G}\right)_t \;=\; r_A\,\alpha \;=\; \left(\dfrac{a_G}{r}\right)$

The acceleration diagram is shown in Figure 6.48, where once $(a_{A/G})_t$ and $(a_{A/G})_n$ are calculate, $\bar{a}_A$ can be determined.

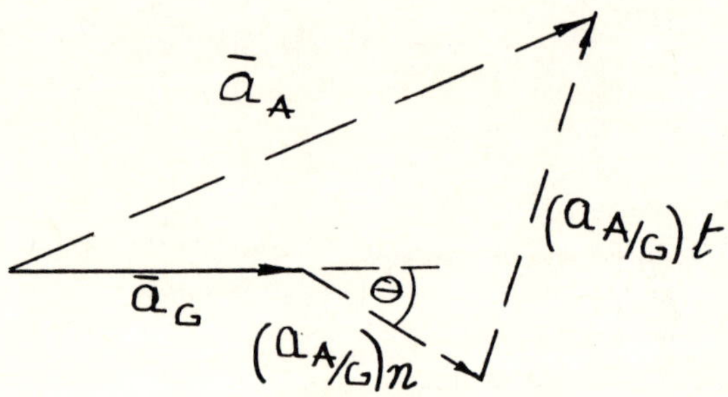

Figure 6.48

If the 'n' and 't' components of $(\bar{a}_A)$ are used, a_A can be found by Pythagoras Theorem, as follows:

$$a_A = \sqrt{\left[\left(a_A\right)_n^2 + \left(a_A\right)_t^2\right]}$$

$$= \sqrt{\left\{\left[a_G\cos\theta + \left(a_{A/G}\right)_n\right]^2 + \left[a_G\sin\theta + \left(a_{A/G}\right)_t\right]^2\right\}} \tag{6.45}$$

$$= \sqrt{\left[\left(r\,\alpha\cos\theta + r_A\,\omega^2\right)^2 + \left(r\,\alpha\sin\theta + r_A\,\alpha\right)^2\right]}$$

where

a_G = acceleration of the centre of gravity
$a_{A/G}$ = rotational acceleration of A, relative to G

Similarly the direction of $\bar{a}_A$ can be calculated.

To Calculate $\bar{a}_C$

$\bar{a}_C$ = the acceleration at the instantaneous centre of zero velocity, namely point 'C'

$$\overline{a_C} = \overline{a_G} + \overline{a_{C/G}} \tag{6.46}$$

The 'n' and 't' components of the term $a_{C/G}$ are given by:

$$(a_{C/G})_n = r\,\omega^2$$

$$(a_{C/G})_t = r\,\alpha$$

These are shown in the acceleration vector diagram of Figure 6.49.

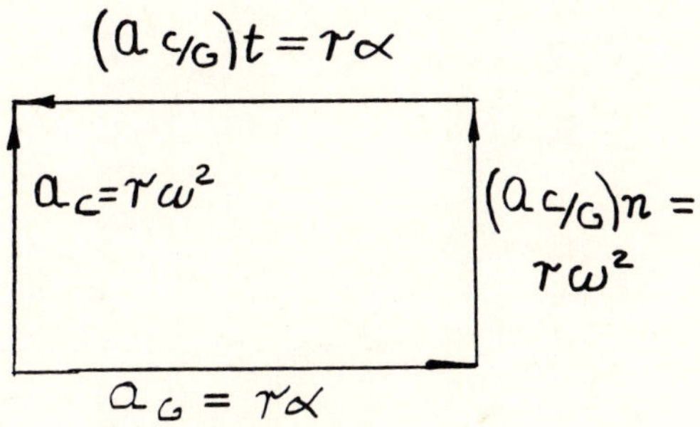

Figure 6.49

From Figure 6.49, $a_C = r\,\omega^2$

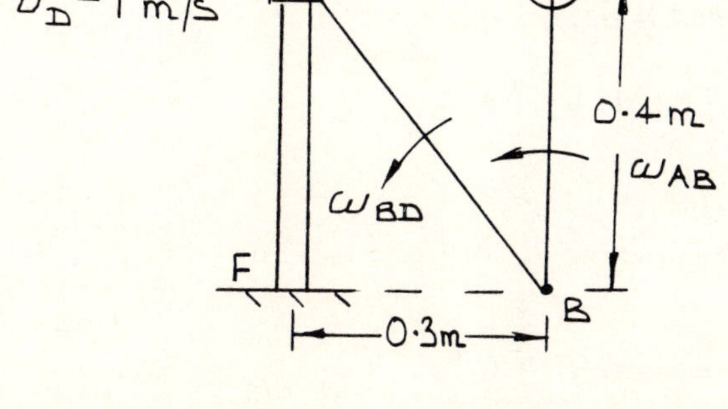

Figure 6.50

Example 6.11

In the mechanism shown in Figure 6.50, the collar 'D' slides vertically downwards on the rod EF. This motion causes the links AB and BD to rotate in a counter-clockwise direction, each with an angular velocity of 7 rad/s. Determine the angular acceleration of these two links at this moment in time. The rod EF may be assumed to be fixed at the points E and F and the collar 'D' is pinned to the link BD.

Consider Link AB

$$\overline{a_B} = \overline{\alpha_{AB}} \times \overline{r_{B/A}} - \omega_{AB}^2 \, r_{B/A}$$

$$= \left(\alpha_{A/B} \, \bar{k}\right) \times \left(-0.4\bar{j}\right) - 7^2 \left(-0.4\bar{j}\right) \qquad (6.47)$$

$$\overline{a_B} = 0.4 \, \alpha_{AB} \, \bar{i} + 19.6 \, \bar{j}$$

Consider Link BD

$$\overline{a_B} = \overline{a_D} + \overline{a_{B/D}}$$

$$= \overline{a_D} + \left(a_{B/D}\right)_n + \left(a_{B/D}\right)_t \qquad (6.48)$$

$$= \overline{a_D} + \overline{\alpha_{B/D}} \times \overline{r_{B/D}} - \omega_{B/D}^2 \, \overline{r_{B/D}}$$

but $\bar{a}_D = -0.5 \, j$

Equating (6.47) and (6.48), we get

$$0.4 \, \alpha_{A/B} \, \bar{i} + 19.6 \, \bar{j}$$

$$= -0.5\bar{j} \times \left(\alpha_{B/D} \, \bar{k}\right) \times \left(0.3\bar{i} - 0.4\bar{j}\right) - 7^2 \left(0.3\bar{i} - 0.4\bar{j}\right)$$

$$= 0.5\bar{j} + 0.3 \, \alpha_{B/D} \, \bar{j} + 0.4 \, \alpha_{B/D} \, \bar{i} - 14.7\bar{i} + 19.6\bar{j}$$

$$= \left(-14.7 + 0.4 \, \alpha_{B/D}\right) \bar{i} + \left(19.1 + 0.3 \, \alpha_{B/D}\right) \bar{j}$$

Equating i's, we get

$$0.4 \, \alpha_{A/B} = -14.7 + 0.4 \, \alpha_{B/D} \qquad (6.49)$$

Equating j's, we get

$$19.6 = 19.1 + 0.3\alpha_{B/D}$$

$$\text{or } \alpha_{B/D} = \frac{0.5}{0.3} \qquad (6.50)$$

$$\alpha_{B/D} = 1.667 \text{ rad/s}^2 \text{ (ccw)}$$

where ccw = counter clockwise

Substituting equation (6.50) into equation (6.49), we get

$$0.4\ \alpha_{AB} = -14.7 + 0.4 \times 1.667$$

$$\text{or } \alpha_{AB} = -35.08 \text{ rad/s}^2 \text{ (cw)}$$

where cw = clockwise

Problems 6.5

1. Determine the angular acceleration of the link AB and the acceleration of the point 'B' of question No. 5 of Problems 6.2, assuming that the point 'A' is travelling to the right with a uniform velocity of 5 m/s.

   [α_{BA} = 77.0 rad/s^2 (ccw); a_B = -91 m/s (downwards)]

2. A ladder AB rests against a wall, as shown in Figure 6.51, when suddenly its base 'A' moves horizontally with a velocity of 2 m/s and an acceleration of 1.5 m/s^2. Determine the angular acceleration of the ladder and its acceleration at its end 'B'.

   [α_{BA} = -1.27 rad/s^2 (cw); a_B = -7.93 m/s^2 (downwards)]

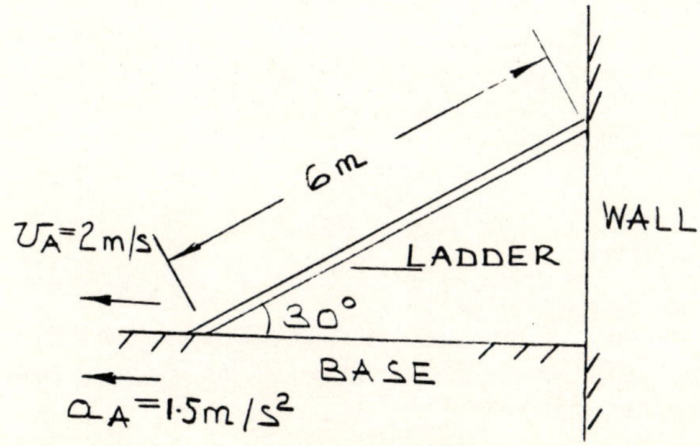

Figure 6.51

3. In the slider-crank mechanism of Figure 6.52, the crank OA rotates clockwise with a constant angular velocity of 3 rad/s. Determine the angular acceleration of the conrod AB and the acceleration of the position 'B'.

[α_{AB} = -0.577 rad/s (cw); a_B = -0.2 m/s^2 (to the left)]

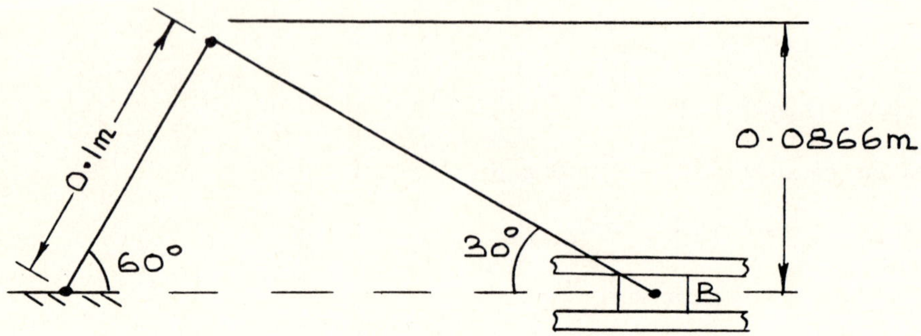

Figure 6.52

4. A rectangular plate is connected to two links at the points 'B' and 'D', as shown in Figure 6.33. If the link AB rotates with a uniform velocity of 3 rad/s about the fixed point 'A', determine the angular acceleration of the plate and the link ED.

[α_{BD} = -3.19 rad/s^2 (cw); α_{ED} = -8.364 rad/s^2 (cw)]

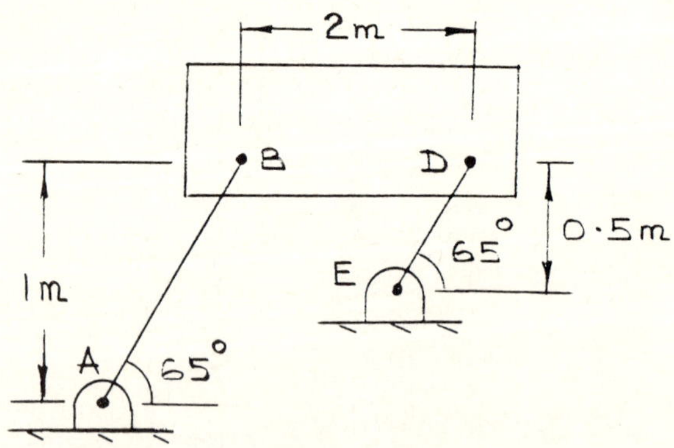

Figure 6.53

5. In the linkage shown in Figure 6.54, the link DB is subjected to a constant angular velocity of 3 rad/s. Determine the angular acceleration for the links OA and AB for this instance in time.

[α_{OA} = 5.06 rad/s^2 (ccw); α_{AB} = -2.496 rad/s^2 (cw)]

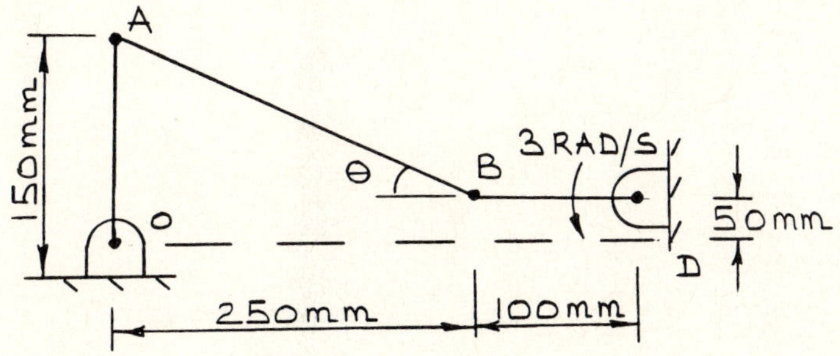

Figure 6.54

6. Determine the angular accelerations of the links AB and BD of Question No. 4 from Problems 6.4, assuming that the link ED is rotating with a constant angular velocity of 5 rad/s.

[α_{AB} = -15.577 rad/s^2 (cw); α_{BD} = 14.537 rad/s^2 (ccw)]

6.9 MOTION RELATIVE TO ROTATING AXES

In the previous sections we considered motion with respect to non-rotating but translating axes. However, for certain problems, particularly where a link is undergoing telescopic action, it may be preferable to consider another system of axes which translates and rotates, as shown in Figure 6.55.

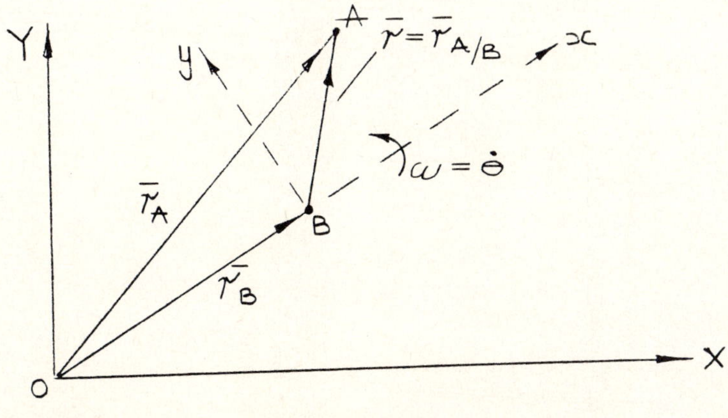

Figure 6.55

In Figure 6.55, the X - Y axes are the fixed reference axes and the x - y axes are the translating and rotating axes, where

ω = the angular velocity of the x-y axes

$$\text{Now } \overline{r_A} = \overline{r_B} + \overline{r_{A/B}} = \overline{r_B} + \bar{r} = r_B + (x\,\bar{\imath} + y\,\bar{\jmath}) \tag{6.51}$$

where the unit vector $\bar{\imath}$ and $\bar{\jmath}$ are in the x and y directions, respectively.

Now the expressions for $\dfrac{d\bar{\imath}}{dt}$ and $\dfrac{d\bar{\jmath}}{dt}$ can be obtained with the aid of

Section 2.6 and Figure 6.56

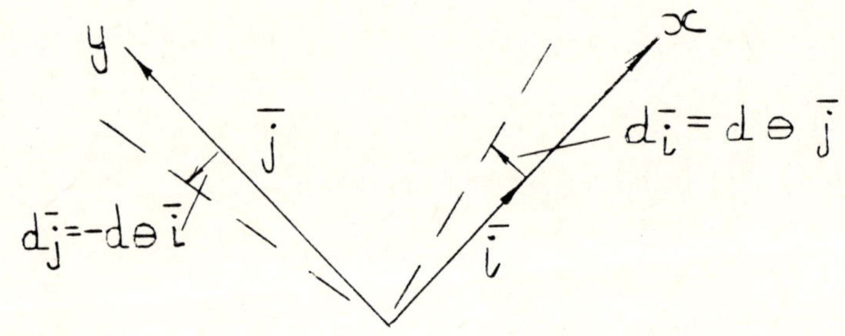

Figure 6.56

From Section 2.6 and Figure 6.56

$$\frac{d\bar{\imath}}{dt} = \frac{d\theta}{dt} \cdot \bar{\jmath} = \omega\bar{\jmath}$$

$$\text{and } \frac{d\bar{\jmath}}{dt} = \frac{-d\theta}{dt} \cdot \bar{\imath} = -\omega\bar{\imath}$$

or in vector terms

$$\frac{d\bar{i}}{dt} = \bar{\omega} \times \bar{i} \tag{6.53}$$

and $$\frac{d\bar{j}}{dt} = \bar{\omega} \times \bar{j} \tag{6.54}$$

because from Figure 6.57

$\bar{\omega} \times \bar{i} = \omega\bar{j}$ and $\bar{\omega} \times \bar{j} = -\omega\bar{i}$

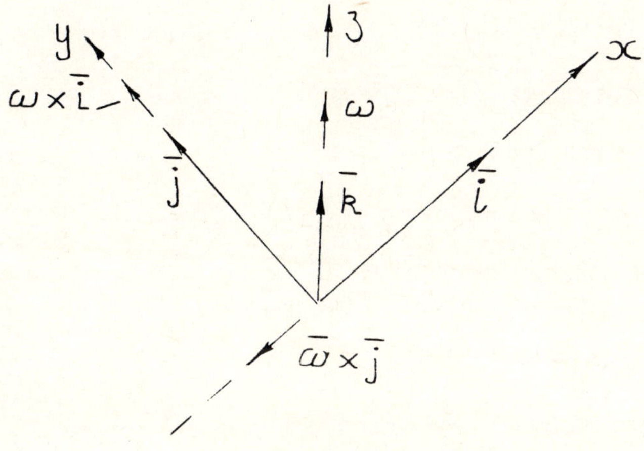

Figure 6.57

6.9.1 Relative Velocity

Differentiating equation (6.51), we get

$$\begin{aligned} \dot{\bar{r}}_A &= \dot{\bar{r}}_B + \frac{d}{dt}\left(x\,\bar{i} + y\,\bar{j}\right) \\ &= \dot{\bar{r}}_B + \left(x\,\frac{d\bar{i}}{dt} + y\,\frac{d\bar{j}}{dt}\right) + \left(\frac{dx}{dt}\,.\,\bar{i} + \frac{dy}{dt}\,.\,\bar{j}\right) \end{aligned} \tag{6.55}$$

$$\text{but } x\frac{d\bar{i}}{dt} + y\frac{d\bar{j}}{dt} = \bar{\omega} \times x\bar{i} + \bar{\omega} \times y\bar{j}$$

$$= \bar{\omega} \times (x\bar{i} + y\bar{j})$$

$$= \bar{\omega} \times \bar{r}$$

Now since the $\frac{dx}{dt}$ and $\frac{dy}{dt}$ are measured the $x - y$ co-ordinate system

$\frac{dx}{dt}\,.\,\bar{i} + \frac{dy}{dt}\,.\,\bar{j} = \overline{v_{rel}}$, so that (6.55) becomes

$$\overline{v_A} = \overline{v_B} + \bar{\omega} \times \overline{r_{A/B}} + \overline{v_{rel}}, \tag{6.56}$$

where

$\bar{v}_A$ = absolute velocity of ‘A’ from the x - y co-ordinate system

$\bar{v}_B$ = absolute velocity of ‘B’ from the x - y co-ordinate system

$\bar{r}_{A/B}$ = position of ‘A’ relative to ‘B’

ω = angular velocity of the x - y co-ordinate system

$\bar{v}_{rel}$ = relative velocity of ‘A’ with respect to ‘B’ (or the relative velocity of ‘A’ as observed from the x - y co-ordinate system)

6.9.2 Relative Acceleration

Differentiating equation (6.56) with respect to time, we get

$$\overline{a_A} = \overline{a_B} + \frac{d\bar{\omega}}{dt} \times \overline{r_{A/B}} + \bar{\omega} \times \frac{d\overline{r_{A/B}}}{dt} + \frac{d\overline{v_{rel}}}{dt} \tag{6.57}$$

Now

$$\frac{d\overline{r_{A/B}}}{dt} = \frac{d}{dt}(x\bar{i} + y\bar{j})$$

$$= \left(x\frac{d\bar{i}}{dt} + y\frac{d\bar{j}}{dt}\right) + \left(\frac{dx}{dt}\,.\,\bar{i} + \frac{dy}{dt}\,.\,\bar{j}\right)$$

$$= \bar{\omega} \times \overline{r_{A/B}} + \overline{v_{rel}}$$

$$\therefore \; \overline{\omega} \; x \; \frac{d\overline{r_{A/B}}}{dt} \;=\; \overline{\omega} \; x \left(\overline{\omega} \; x \; \overline{r_{A/B}} + \overline{v_{rel}} \right)$$

$$= \overline{\omega} \; x \left(\overline{\omega} \; x \; \overline{r_{A/B}} \right) + \overline{\omega} \; x \; \overline{v_{rel}} \qquad (6.58)$$

Similarly

$$\frac{dv_{rel}}{dt} = \frac{d}{dt}\left(\frac{dy}{dt} \, . \, \overline{i} + \frac{dy}{dt} \, . \, \overline{j} \right)$$

$$= \left(\frac{dx}{dt} \, . \, \frac{d\overline{i}}{dt} + \frac{dy}{dt} \, . \, \frac{d\overline{j}}{dt} \right) + \left(\frac{d^2x}{dt^2} \, . \, \overline{i} + \frac{d^2y}{dt^2} \, . \, \overline{j} \right) \qquad (6.59)$$

$$= \overline{\omega} \; x \; \overline{v_{rel}} + \overline{a_{rel}}$$

Substituting equations (6.58) and (6.59) into equation (6.57), we get

$$\overline{a_A} = \overline{a_B} + \frac{d\overline{\omega}}{dt} \; x \; \overline{r_{A/B}} + \overline{\omega} \; x \left(\overline{\omega} \; x \; \overline{r_{A/B}} \right) + 2\overline{\omega} \; x \; \overline{v_{rel}} + \overline{a_{rel}} \qquad (6.59)$$

where

$\overline{a_A}$ = accleration of `A′ with respect to the $x - y$ axes

$\overline{a_B}$ = acceleration of `B′ with respect to the $x - y$ axes

$\overline{\omega}$ = the angular velocity of the $x - y$ axes

$\frac{d\overline{\omega}}{dt} = \overline{\alpha}$ = the angular acceleration of the $x - y$ axes

Example 6.12

A rod OA is rotating about the point ‘O’ with an angular velocity of 2.5 rad/s, together with an angular acceleration of 1.5 rad/s^2, as shown in Figure 6.58. If while this rod is rotating, a sleeve ‘G’ is travelling towards ‘A’ on the rod in an axial direction, with a relative velocity of 3.5 m/s, together with a relative acceleration of 2 m/s^2, determine the velocity and acceleration of the sleeve and its Coriolis acceleration.

It is convenient to choose the X - Y and the x - y co-ordinate systems shown in Figure 6.58, so that $\overline{v}_{rel}$, $\overline{a}_{rel}$, ω and $\overline{\alpha}$ can be obtained by inspection.

At 'O', $v_o = a_o = 0$

By inspection

$\bar{v}_{rel} = 3.5\,\bar{i}$ m/s,

$\bar{a}_{rel} = 2\,\bar{i}$ m/s^2,

$\bar{\omega}$ = angular velocity of the x - y axes

$= -2.5\,\bar{k}$

$\bar{\alpha}$ = angular acceleration of the x - y axes

$= -1.5\,\bar{k}$

$\bar{r}_{c/o} = 0.5$ i

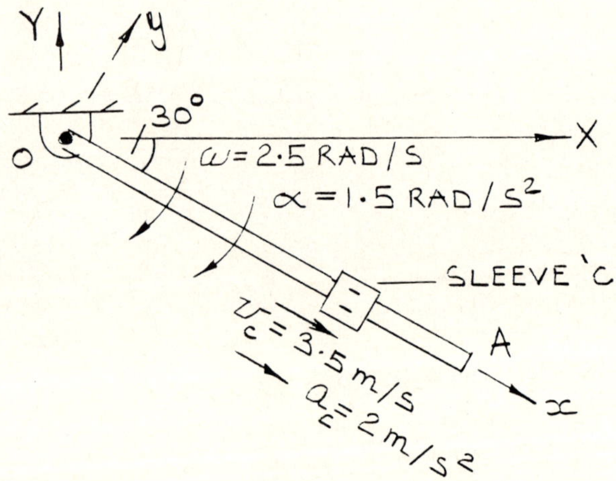

Figure 6.58

From equation (6.56),

$$\overline{v_c} = \overline{v_o} + \overline{\omega} \times \overline{r_{c/o}} + \overline{v_{rel}}$$

$$= 0 - 2.5\bar{k} \times (0.5\bar{i}) + 3.5\bar{i}$$

$$= -1.25\bar{j} + 3.5\bar{i},$$

$$\text{or } \overline{v_c} = (3.5\bar{i} - 1.25\bar{j})\ m/s$$

From equation (6.60)

$$\overline{a_c} = \overline{a_o} + \overline{\alpha} \times \overline{r_{c/o}} + \overline{\omega} \times \left(\overline{\omega} \times \overline{r_{c/o}}\right) + 2\overline{\omega} \times \overline{v_{rel}} + \overline{a_{rel}}$$

$$= 0 - 1.5\overline{k} \times \left(0.5\overline{i}\right) - 2.5\overline{k} \times \left(-2.5\overline{k} \times 0.5\overline{i}\right) - 2 \times 2.5\overline{k} \times 3.5\overline{i} + 2\overline{i}$$

$$= -0.75\overline{j} - 2.5\overline{k} \times \left(-1.25\overline{j}\right) - 17.5\overline{j} + 2\overline{i}$$

$$= -0.75\overline{j} - 3.125\overline{i} - 17.5\overline{j} + 2\overline{i}$$

$$\text{or } \overline{a_c} = \left(-1.25\overline{i} - 18.25\overline{j}\right) \text{ m/s}^2$$

The Coriolis component of acceleration is given by:

$$\overline{a}\,(Coriolis) = 2\overline{\omega} \times \overline{v_{rel}}$$

$$= -2 \times 2.5\overline{k} \times 2.5\overline{i} \tag{6.61}$$

$$= \left(-17.5\overline{j}\right) \text{ m/s}^2$$

Example 6.13

In the mechanism shown in Figure 6.59, the rod OA rotates about the fixed point 'O', with an angular velocity of 2.5 rad/s and an angular acceleration of 3.5 rad/s^2 and this causes the sleeve 'B' to move axially along the link CD. If the sleeve 'B' is pin-connected to the link OA, determine the angular velocity and angular acceleration of the link CD.

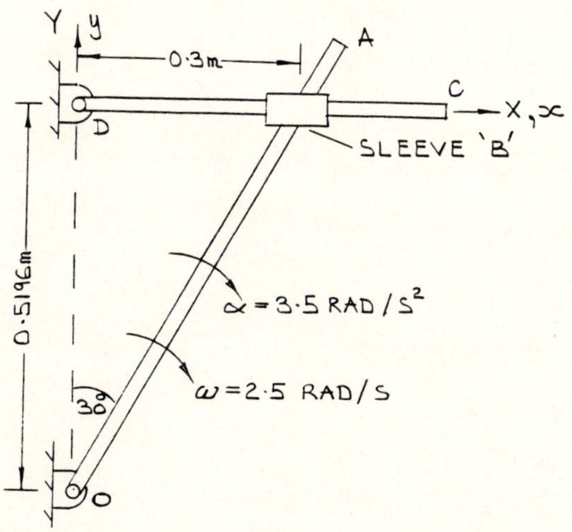

Figure 6.59

Co-ordinates

As the sleeve moves along the link CD, it will be convenient to make the x - y co-ordinate system as shown in Figure 6.59, so that $\bar{v}_{rel}$ and $\bar{a}_{rel}$ can be easily determined.

At the point 'D'

$\bar{v}_D = \bar{a}_D = 0$

$\omega_{B/D}$ = angular velocity of the x - y axes

$\bar{\alpha}_{B/D}$ = angular acceleration of the x - y axes

From equation (6.56)

$$\overline{v_B} = \overline{v_D} + \overline{\omega_{B/D}} \times \overline{r_{B/D}} + \left(\overline{v_{B/D}}\right)_{rel} \tag{6.62}$$

Now the sleeve 'B' moves along a circular path, because the link OA rotates about 'O', so that

$$\begin{aligned} \overline{v_B} &= \text{velocity of `B'} = \overline{\omega_{O/A}} \times \overline{r_{O/A}} \\ &= -2.5\bar{k} \times \left(0.3\bar{i} + 0.5196\bar{j}\right) \\ &= \left(-0.75\bar{j} + 1.3\bar{i}\right) \end{aligned} \tag{6.63}$$

or $\overline{v_B} = \left(1.3\bar{i} - 0.75\bar{j}\right)$ m/s

From equation (6.60)

$$\overline{a_B} = \overline{a_D} + \overline{\alpha_{B/D}} \times \overline{r_{B/D}} + \overline{\omega_{B/D}} \times \left(\overline{\omega_{B/D}} \times \overline{r_{B/D}}\right) + 2\overline{\omega_{B/D}} \times \left(\overline{v_{B/D}}\right)_{rel} + \left(\overline{a_{B/D}}\right)_{rel} \tag{6.64}$$

Now

$$\begin{aligned} \overline{a_B} &= \overline{\alpha_{O/B}} \times \overline{r_{O/B}} + \overline{\omega_{O/B}} \times \left(\overline{\omega_{O/B}} \times \overline{r_{O/B}}\right) \\ &= -3.5\bar{k} \times \left(0.3\bar{i} + 0.5196\bar{j}\right) - 2.5\bar{k} \times \left[-2.5\bar{k} \times \left(0.3\bar{i} + 0.5196\bar{j}\right)\right] \\ &= -1.05\bar{j} + 1.819\bar{i} - 2.5\bar{k} \times \left(-0.75\bar{j} + 1.3\bar{i}\right) \end{aligned}$$

$$\text{or } \overline{a_B} = \left(-0.056\bar{i} - 4.3\bar{j}\right) \text{ m/s}^2 \tag{6.65}$$

Equating (6.62) and (6.63), we get

$$1.3\vec{i} - 0.75\vec{j} = \omega_{B/D} \, . \, \vec{k} \times (0.3\vec{i}) + \left(\overline{v_{B/D}}\right)_{rel}$$

$$= \omega_{B/D} \, . \, \vec{j} + \left(\overline{v_{B/D}}\right)_{rel} \, . \, \vec{i}$$

Equating j's

$$\omega_{B/D} = -0.75 \text{ rad/s}$$

Equating i's

$$\left(v_{B/D}\right)_{rel} = 1.3 \text{ m/s}$$

Equating (6.64) and (6.65), we get

$$-0.056\vec{i} - 4.3\vec{j} = 0 + \overline{\alpha_{B/D}} \times \overline{r_{B/D}} + \overline{\omega_{B/D}} \times \left[\overline{\omega_{B/D}} \times (0.3\vec{i})\right]$$

$$+ 2\overline{\omega_{B/D}} \times \left(\overline{v_{B/D}}\right)_{rel} + \left(\overline{a_{B/D}}\right)_{rel} = \alpha_{B/D} \, . \, \vec{k} \times (0.3\vec{i}) + w_{B/D}\vec{k} \times \left[\, \omega_{B/D}\vec{k} \times (0.3\vec{i}) \,\right]$$

$$+ 2\,\omega_{B/D}\,\vec{k} \times (1.3\vec{i}) + a_{B/D_{rel}}\,\vec{i} = \alpha_{B/D} \, . \, 0.3\vec{j} + \omega_{B/D} \, . \, \vec{k}\left(0.3\,\omega_{B/D} \, . \, \vec{j} + 2.6\omega_{B/D} \, . \, \vec{j} + \left(a_{B/D}\right)\right)_{rel} . \, \vec{i}$$

$$\text{or } 0.056i - 4.3j = \alpha_{B/D} \, . \, 0.3\vec{j} - \omega_{B/D} \, . \, 0.3\vec{i} + 2.6\omega_{B/D}\vec{j} + \left(a_{B/D}\right)_{rel} . \, \vec{i}$$

Equating j's

$$-4.3 = 0.3\,\alpha_{B/D} + 2.6\,\omega_{B/D}$$

$$\text{or } \alpha_{B/D} = (-4.3 + 2.6 \times 0.75)/0.3$$

$$= -7.83 \text{ rad/s}^2$$

Equating i's

$$-0.056 = -0.3\,\omega_{B/D} + \left(a_{B/D}\right)_{rel}$$

$$\therefore \left(a_{B/D}\right)_{rel} = -0.281 \text{ m/s}^2$$

Problems 6.6

1. An 'L' shaped tube rotates about 'O' in a clockwise direction, looking upwards. The angular velocity of AB is 3.5 rad/s and its angular acceleration is 4.5 rad/s^2, looking upwards. At the same time that this occurs, a ball of negligible size, moves radially outwards with a relative velocity of 2 m/s and a relative acceleration of 1.5 m/s^2. Determine the velocity and acceleration of the ball.

[$v_B = (2\bar{i} + 3.5\bar{j})$ m/s; $a_B = (-10.75\bar{i} + 18.5\bar{j})$ m/s^2]

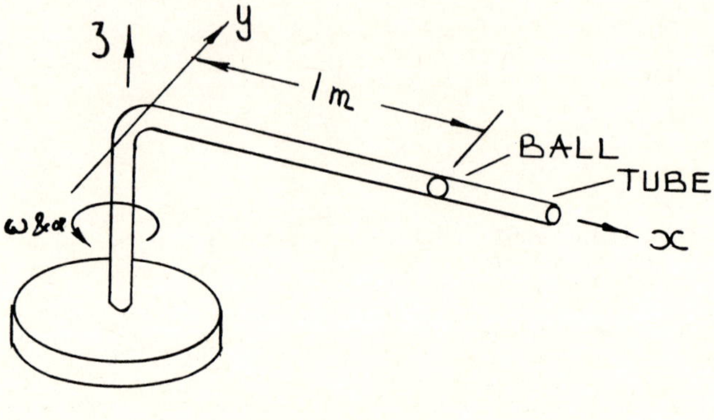

Figure 6.60

2. At a certain set of crossroads, car 'A' is approaching car 'B' at the crossroads with a constant speed of 50 km/h, while car B is travelling around a bend of radius 100 m, with a constant speed of 70 km/h.

Determine the relative velocity to an observer in car 'B' that car 'A' appears to approach the observer.

[$\bar{v}_{rel} = (13.89\bar{i} - 23.68\bar{j})$ m/s]

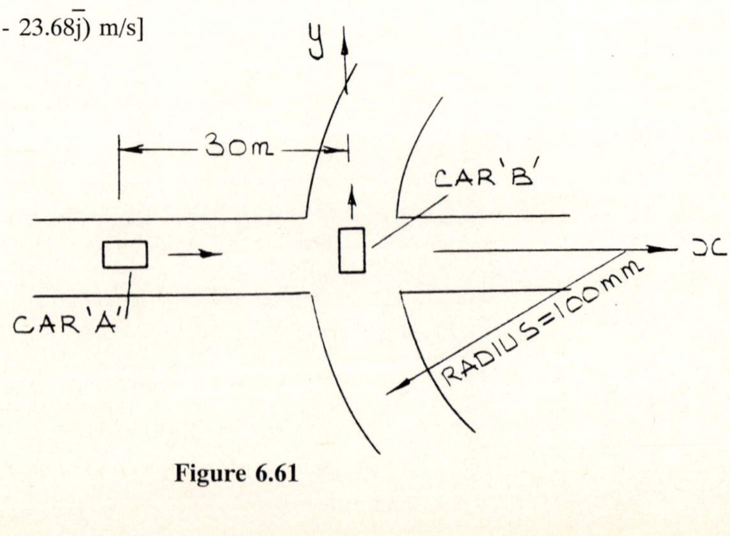

Figure 6.61

3. A racing circuit is such that at a certain point two sections of the circuit are near each other, as shown in Figure 6.62. If both cars are rounding their respective bends with constant speeds of 100 km/h, determine the relative velocity that car 'B' appears to have to the driver of car 'A'.

[v_{rel} = -33.32i m/s]

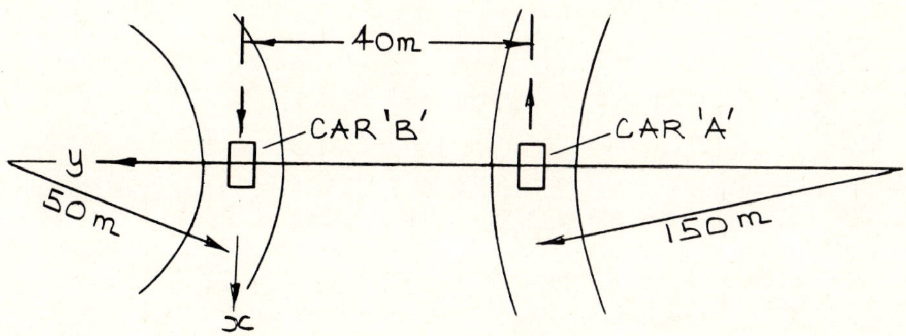

Figure 6.62

4. Two aircraft are flying at the same altitude in the directions shown in Figure 6.63. Aircraft 'A' has a speed of 800 km/h and an acceleration of 40 km/h^2, while aircraft 'B' has a speed of 700 km/h and a deceleration of 60 km/h^2.

Determine the apparent velocity and acceleration that plane 'B' appears to have to an observer in plane 'A'.

[$\bar{v}_{rel}$ = $93\bar{j}$ m/s; $\bar{a}_{rel}$ = $(-1550\bar{i} + 100.6\bar{j})$ m/s^2]

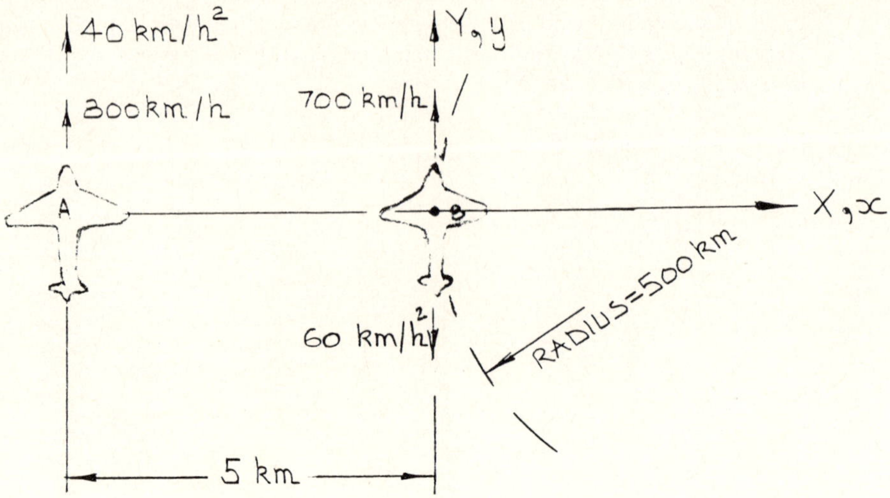

Figure 6.63

5. A mechanism is shown in Figure 6.64, where the rod BC rotates about the point 'B', in a clockwise direction, with an angular velocity of 3.5 rad/s and an angular acceleration of 2.5 rad/s^2. At its end 'C', the rod BC has pinned to it a sleeve 'C', which can move axially along the rod AD. Determine the angular velocity and angular acceleration of the rod AD in the position shown.

[$\omega_{C/D}$ = -3.5 rad/s (cw); $\alpha_{C/D}$ = 6.88 rad/s^2 (ccw)]

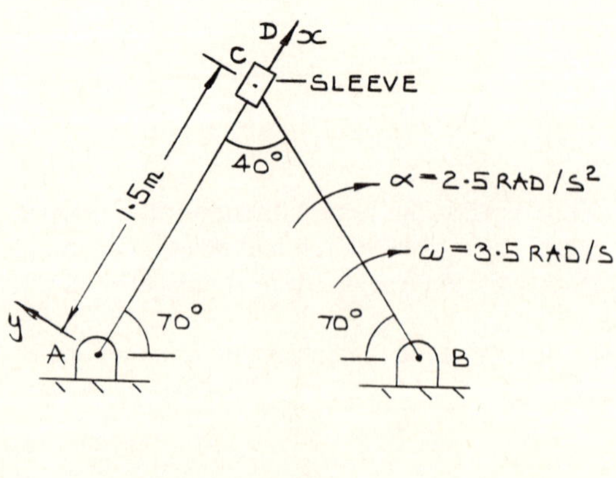

Figure 6.64

6. A mechanism is shown in Figure 6.65, where the slotted link AB rotates about the point 'A' with a constant angular velocity of 2.5 rad/s. Additionally, the link DB rotates about the point 'D' and the end 'B' is confined to move in the slot of the link AB. Determine the angular velocity and angular acceleration of the pin 'C' relative to the rotating slot.

[$\omega_{C/D}$ = -10 rad/s (cw); v_{rel} = -1.732 m/s;
$\alpha_{C/D}$ = 259.8 rad/s^2 (ccw); a_{rel} = 57.497 m/s^2]

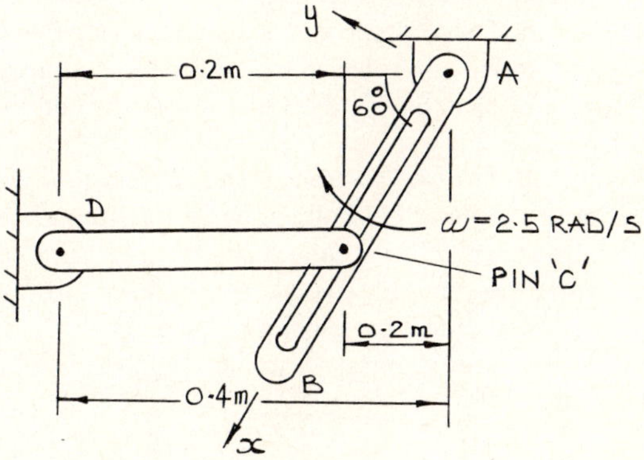

Figure 6.65

CHAPTER SEVEN

Kinetics of a Rigid Body

7.1 INTRODUCTION

In this Chapter, we will commence by defining a term call **mass moment of inertia**. This term is important in the kinetics of rigid bodies, as a study of this topic involves the forces and moments causing translation and rotation of rigid bodies. It is the magnitude of the mass moment of inertia that is related to the rigid body's ability to continue or resist rotary motion.

The Chapter will be divided into four sections. The first section will be on determining the mass moment of inertia of various rigid bodies, while the second section will solve kinetics problems using force: mass: acceleration relationships. The third section will solve kinetics problems involving work, power and energy while the fourth section will solve these problems using impulse and momentum considerations. In the last three sections it will be shown how different methods lend themselves to the satisfactory solution of certain problems.

7.2 MASS MOMENT OF INERTIA 'I'

The mass moment of inertia, denoted by the symbol 'I', is used in the rotational equivalent of the mass: force: acceleration relationship, as shown by equation (7.1)

$$\overline{M} = I\,\overline{\alpha} \tag{7.1}$$

where

$\overline{M}$ = moment

I = mass moment of inertia

$\overline{\alpha}$ = angular acceleration

I = $\Sigma\, \mathrm{d}m \,.\, r^2$

dm = element of mass in kg

r = distance of the element dm from the point which I is being calculated for the rigid body

The units of 'I' are $\mathrm{kg.m^2}$

For a continuous rigid body of density 'ρ_m'

I = $\int r^2 \,.\, \rho_m\, \mathrm{d(vol)}$

If 'ρ_m' is a constant

$$I = \rho_m \int r^2 \, d(vol) \qquad (7.2)$$

$$= mk^2$$

where m = mass of body
k = radius of gyration of the body about a point where 'I' is being calculated.

Example 7.1

Calculate 'I' about the centroid for the solid disc of uniform thickness 't', shown in Figure 7.1.

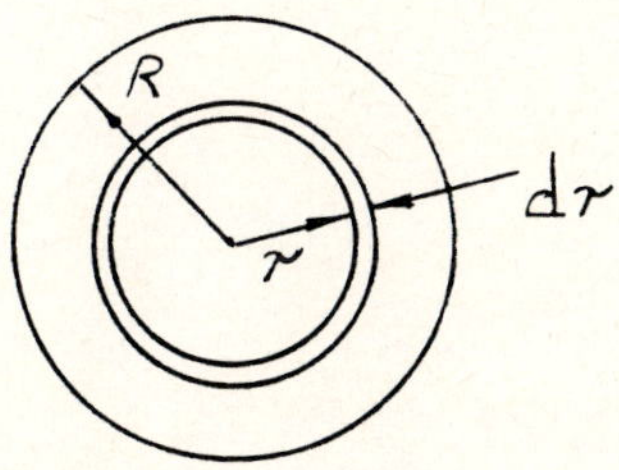

Figure 7.1 Solid Disc

The mass moment of inertia of the annular element about the centroid =

$$d\,I = d\,(vol)\,.\,\rho_m\, r^2$$

$$= \pi\, r\,.\,dr\,.\,t\,.\,\rho_m\, r^2$$

$$\therefore I = 2\pi\, \rho_m\,.\,t \int_o^R r^3\, dr$$

$$= 2\pi\, \rho_m\,.\,t \left[\frac{r^4}{4}\right]_o^r$$

$$I = \pi\, R^4\, \rho_m\,.\,t/2$$

but m = mass of disc = $\pi\, R^2\,.\,t\,\rho_m$

$$\therefore I = m\, R^2/2 \qquad (7.3)$$

Example 7.2

Calculate 'I' for a thin solid road about an end. The rod may be assumed to be of constant cross-sectional area 'a' and of length 'L'.

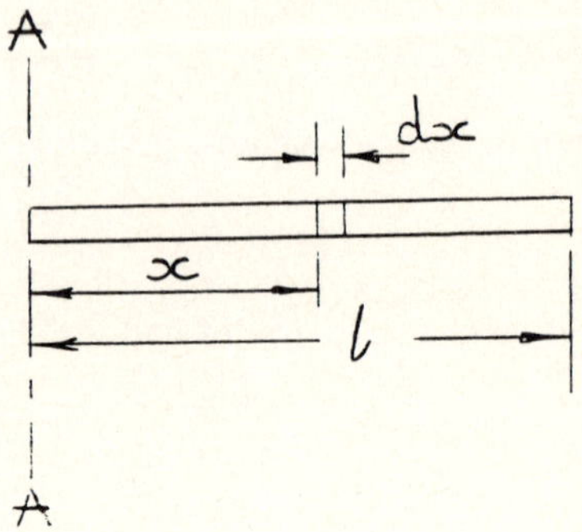

Figure 7.2

The mass moment of inertia of an element of length dx, about the end A - A =

$$\mathrm{d}I = a \,.\, \mathrm{d}x \times \rho_m x^2$$

$$\therefore I = \rho_m a \int_o^L x^2 \,\mathrm{d}x = \rho_m a \left[x^3/3\right]_o^L$$

$$= \rho_m a L^3/3$$

but m = mass of rod = $\rho_m a L$

$$\therefore I = m L^2/3 \text{ (about an end)} \tag{7.4}$$

Example 7.3

Calculate 'I' for the problem of Example 7.2 about mid-length

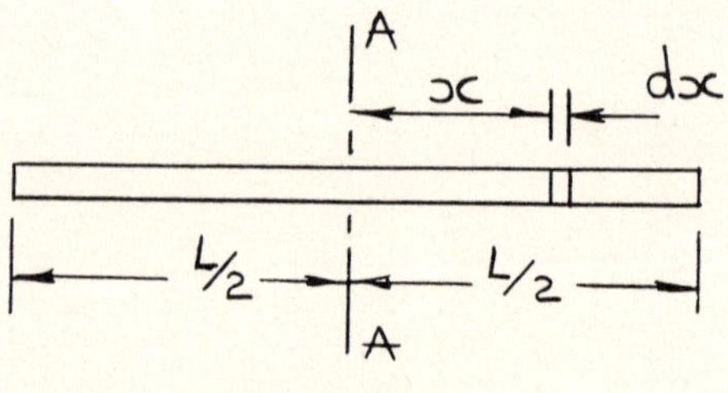

Figure 7.3

$$I = \rho_m\ a \int_{-L/2}^{L/2} x^2\ .\ \mathrm{d}x$$

$$= \rho_m\ a\ \left[x^3/3\right]_o^{L/2}$$

$$= 2\ \rho_m\ a\ \left[x^3/3\right]_o^{L/2}$$

$$= \frac{2\ \rho_m\ a}{3}\ \left[L^3/8\right]$$

$$I = \rho_m\ a\ L^3/12$$

$$= m\ L^2/12 \qquad (7.5)$$

where

m = mass of rod = $\rho_m\ a\ L$

7.2.1 Parallel Axes Theorem

This theorem can be useful if 'I' is known about an axis passing through the centroid of the body and it is required to determine the mass moment of inertia about an axis parallel to the axis passing through the centroid.

Consider an element of mass, namely dm, as shown in Figure 7.4.

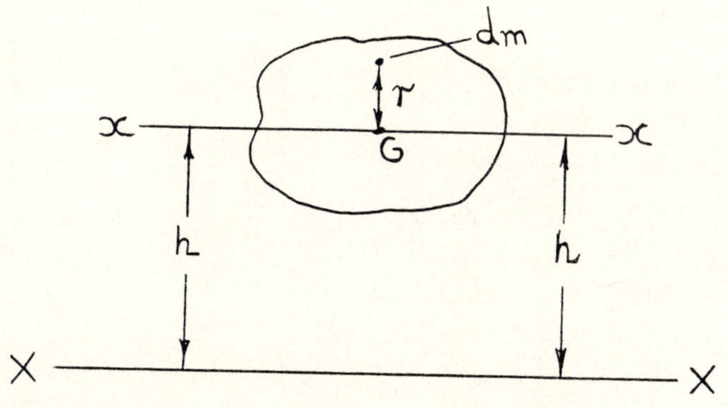

Figure 7.4 Parallel Axes

Let XX be the axis about which the mass moment of inertia is required in terms of the mass moment of inertia about the centroid, namely I_{xx}.

Now

$$I_{xx} = \sum dm \,.\, r^2$$

$$\text{and } I_{XX} = \sum dm \,.\, (h + r)^2$$

$$= \sum dm \,.\, (h^2 + r^2 + 2rh)$$

but Σ dm . 2 r h = 0, because 2 h . dm is a constant and 'r' is from the centroid.

$$\therefore I_{XX} = \sum dm \,(h^2 + r^2)$$

$$= I_{xx} + \sum dm \,.\, h^2 \qquad (7.6)$$

Equation (7.6) gives I_{XX} in terms of I_{xx} and h and is known as the **parallel axes theorem** because xx and XX are in fact parallel axes.

Example 7.4

Using the parallel axes theorem, calculate 'I' about an end for a slender rod, given the mass moment of inertia of a slender rod about its centroid, namely, I_{xx}.

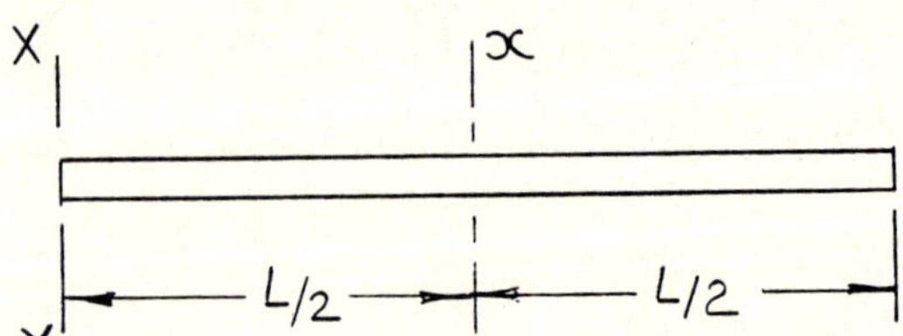

Figure 7.5

From the parallel axes theorem,

$$I_{XX} = I_{xx} + \sum dm \,.\, h^2$$

$$= mL^2/12 + m \times (L/2)^2$$

$$= mL^2\left(\frac{1}{12} + \frac{1}{4}\right)$$

$$I_{XX} = mL^2/3 \text{ as required}$$

Example 7.5

A 'T' shaped pendulum is shown in Figure 7.6, where the mass of each of the two components of the tee is 5 kg. Determine the mass moment of inertia:

(a) about 'O'
(b) about the centroid 'G'.

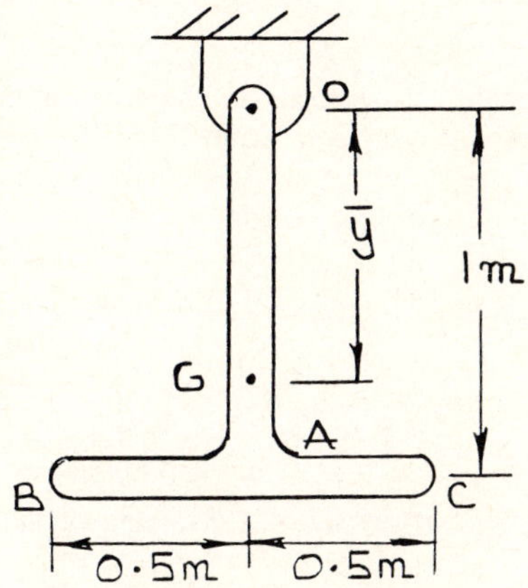

Figure 7.6

Let m = mass of each rod = 5 kg

$\bar{y}$ = distance of the centroid of the tee from 'O'

(a) $(I_{OA})_o = mL^2/3 = 5 \times 1^2/3 = 1.667 \text{ kg . m}^2$

$(I_{BC})_o = mL^2/12 + mL^2$

$= 5 \times 1^2/12 + 5 \times 1^2 = 5.417 \text{ kg . m}^2$

$\therefore I_o = (I_{OA})_o + (I_{BC})_o = 7.084 \text{ kg . m}^2$

(b) To determine $\bar{y}$, take moments about 'O'

$2\,m\bar{y} = m \times 0.5 + m \times 1$

$\therefore \bar{y} = 0.75\,\text{m}$

From equation (7.6)

$I_G = I_o - 2\,m\,\bar{y}^2$

$= 7.084 - 2 \times 5 \times 0.75^2$

$I_G = 1.46$ kg m^2

Problems 7.1

1. Calculate the mass moment of inertia of a sphere of radius 'a' about a diameter.

   [$2\,ma^2/5$]

2. Calculate the mass moment of inertia of a solid disk of radius 'R' about a diameter.

   [$m\,R^2/4$]

3. Calculate the mass moment of inertia of a thin ring of radius 'R' about a diameter.

   [$m\,R^2/2$]

4. Calculate the mass moment of inertia of a thin-walled circular cylinder, of radius 'R' and length 'L' about its mid-length diameter.

   [$m\,R^2/2 + m\,L^2/12$]

5. Calculate the mass moment of inertia of a solid circular cylinder, of radius 'R' and length 'L' about its mid-length diameter.

   [$m\,R^2/4 + m\,L^2/12$]

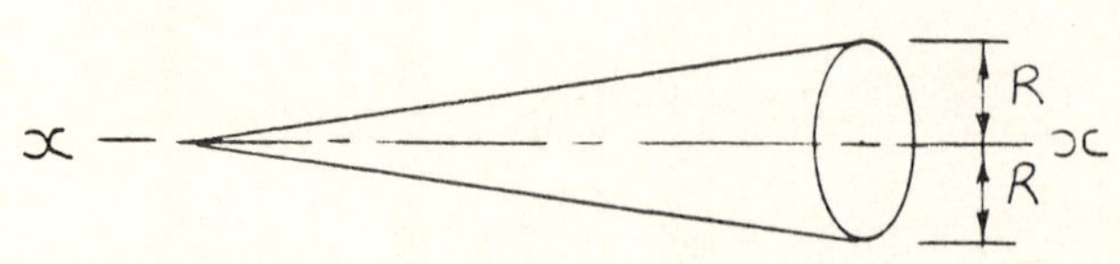

Figure 7.7

6. Calculate the mass moment of inertia of a solid circular cone about its axis of symmetry x - x, as shown in Figure 7.7.

   [$3\ mR^2/10$]

7.3 FORCE: MASS: ACCELERATION

In Chapter 3, we considered the relationship between force, mass and acceleration for a particle. In this section, we extend this theory to rigid body motion, where in addition to translatory motion, we consider rotary motion.

Due to translatory motion, we have from Newton's 2nd Law,

$$\Sigma \bar{F} = m\bar{a}$$

where

$\Sigma \bar{F}$ = the resultant of all the external forces acting on the body
m = mass of the body
$\bar{a}$ = acceleration vector

From Chapter 5, the moment equation about the mass centre 'G' is given by:

$$\Sigma \bar{M}_G = \dot{\bar{H}}_G \qquad (7.8)$$

where $\Sigma \bar{M}_G$ = the resultant of the couples acting on the body
$\dot{\bar{H}}_G$ = the derivative of the angular momentum with respect to time (7.9)

Equations (7.7) and (7.8) are shown in Figure 7.8, where the free body diagram shows all the forces acting on the rigid body and Figure 7.8(b) shows the resultant force and moment vectors on the body. Figure 7.8(c) shows the kinetic diagram due to these forces and couples.

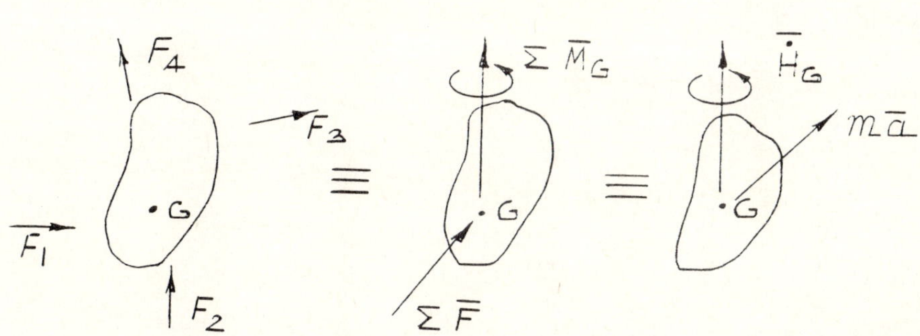

(a) Free Body Diagram **(b) Resultant Force and Couple** **(c) Kinetic Diagram**

Figure 7.8

Consider now the angular momentum about the centroid of a planar body, as shown in Figure 7.9.

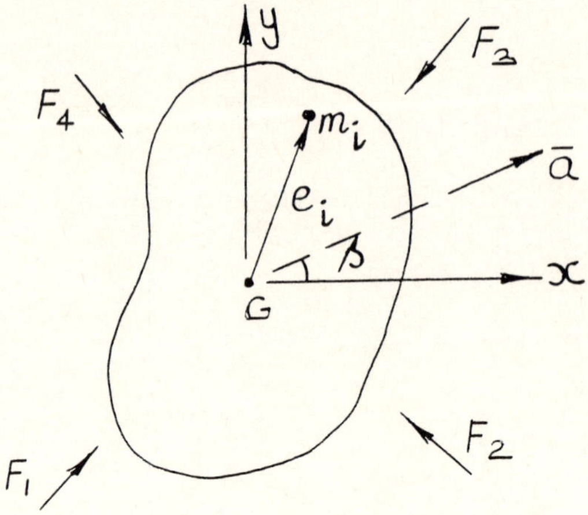

Figure 7.9

The angular momentum

$$\overline{H_G} = \sum \overline{\rho_i} \times m_i \, \dot{\overline{\rho}}_i$$

but

$$\overline{\rho_i} \times \dot{\overline{\rho}} = \rho_i^2 \, \omega, \text{ normal to the x - y plane}$$

Therefore the magnitude of $\bar{H}_G = H_G = \Sigma \, \rho_i^2 \, m_i \, \omega$

$$\text{but} \sum \rho_i^2 \, m_i = I_G \qquad (7.11)$$

$$\therefore H_G = I_G \, \omega \qquad (7.12)$$

$$\text{Now } \dot{H}_G = \sum M_G$$

$$\therefore \sum M_G = I_G \, \dot{\omega} = I_G \, \alpha \qquad (7.13)$$

where

I_G = mass moment of inertia about the centre of gravity 'G' and perpendicular to the plane.

Equations (7.7) and (7.13) can be shown pictorially in Figure 7.10, where figure 7.10(a) is the free body diagram and Figure 7.10(b) is the kinetic diagram.

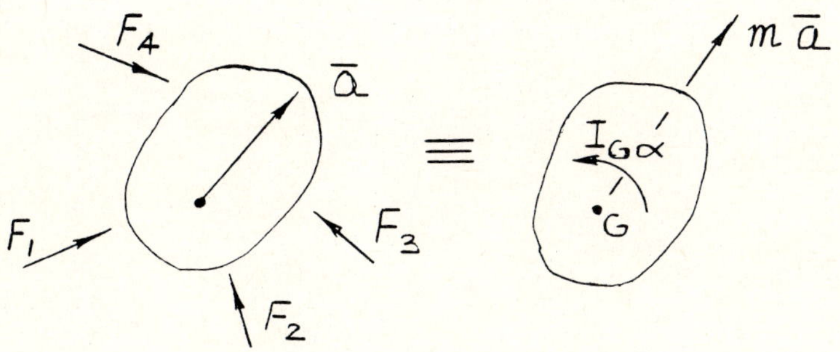

(a) **Free Body Diagram** (b) **Kinetic Diagram**

Figure 7.10

7.3.1 Translation

In this case there is no rotation, so equations (7.7) and (7.13) becomes:

$$\sum F = m\,\bar{a} \tag{7.14}$$

$$\text{and} \sum M_G = 0 \tag{7.15}$$

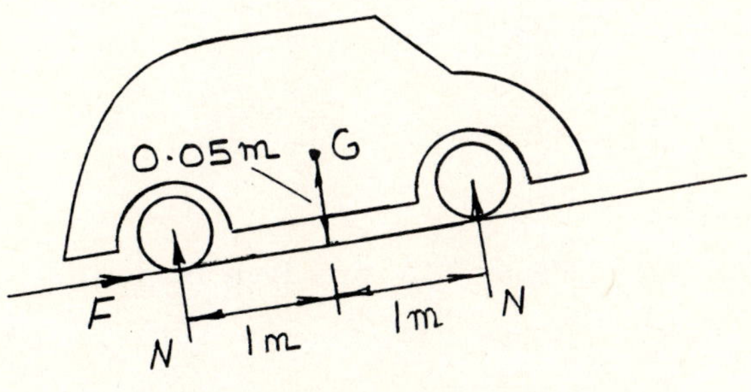

Figure 7.11

Example 7.6

A rear wheeled drive car of mass 1,000 kg, ascends a slope of 7 percent, as shown in Figure 7.11. The coefficient of kinetic friction between the tyres and the road is 0.6. Determine the frictional force on the rear tyres and the normal forces on both front and rear tyres, assuming that the car accelerates uniformly from rest to a speed of 60 km/h in a distance of 70 m.

$$v = \frac{60}{3.6} = 16.667\ \text{m/s}$$

From equation (2.12)

$$v^2 = v_o^2 + 2\,as$$

$$\text{or } 16.667^2 = 0 + 2 \times a \times 70$$

$$\therefore\ a = 1.984\ \text{m/s}^2$$

The free body diagram is shown in Figure 7.12.

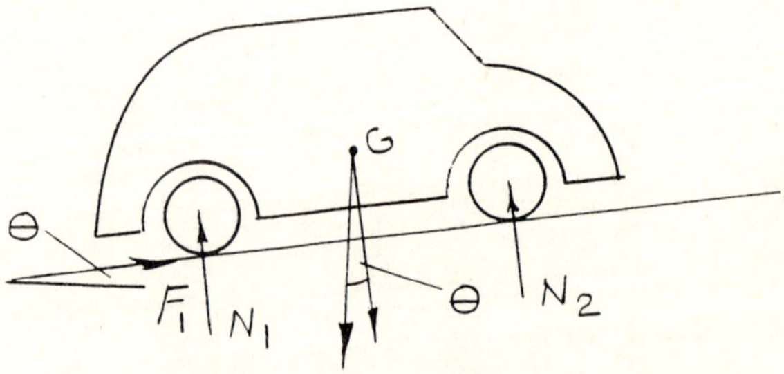

Figure 7.12 Free Body Diagram

$$\theta = \tan^{-1}(7/100) = 4°$$

Applying Newton's 2nd Law up the slope, we get

$$F_1 - mg \sin\theta = ma$$

$$\text{or } F_1 = m\left(a + \sin\theta\right)$$

$$= 1000\left(1.984 + 9.81 \sin 4°\right) \qquad (7.16)$$

$$= 1984 + 684.3$$

$$F_1 = 2668.3 \text{ N}$$

Now

$$\sum M_G = I_G\, \alpha = 0$$

$$\therefore N_1 \text{ x } 1 = N_2 \text{ x } 1 + F_1 \text{ x } 0.5$$

$$\text{or } N_1 = N_2 + 0.5\, F_1 \qquad (7.17)$$

Resolving forces perpendicular to the slope we get

$$N_1 + N_2 = mg \cos\theta$$

$$= 1000 \text{ x } 9.81 \cos 4° \qquad (7.18)$$

$$= 9786.1$$

Substituting equation (7.17) into equation (7.18), we get

$$N_2 + 1334.2 + N_2 = 9786.1$$

$$\text{or } N_2 = 422.6 \text{ N} \qquad (7.19)$$

Substituting equation (7.19) into equation (7.17)

$$N_1 = 5560.2 \text{ N}$$

To check whether or not there is any **slipping** of the rear tyres.

$$F_1 = \mu_s N_1$$

$$\text{or } \mu_s = F_1 / N_1$$

$$= 2668.3 / 5560.2$$

$$\mu_s = 0.48$$

As the required μ_s is less than μ_k, there is **no slipping** of the rear wheels.

Example 7.7

A crate of mass 60 kg rests on a flat but rough horizontal surface, as shown in Figure 7.13. If a horizontal force of 700 N is applied to the crate and the kinetic coefficient of friction between the crate and the surface is 0.3, determine the crate's acceleration.

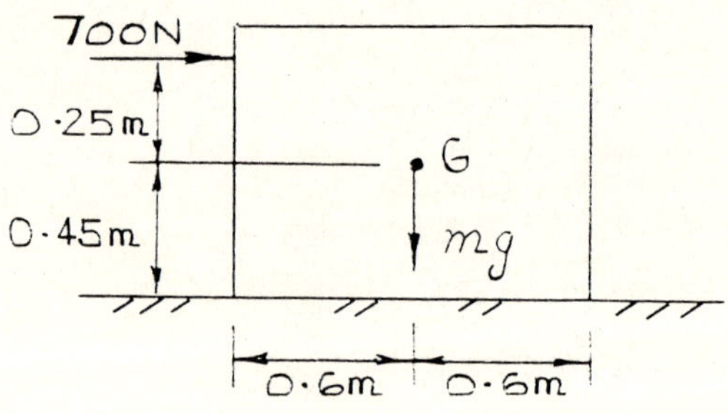

Figure 7.13

The free body diagram is shown in Figure 7.14, where the vertical reaction N_1, is assumed to act at a distance x on the base of the crate and to the right of its centre line. If $x > 0.6$m, the crate will overturn and a different calculation will have to be carried out. Let us assume that $\mu_k = 0.3$ so that the crate slips.

Resolving Forces Vertically

$$N_1 = mg = 60 \times 9.81 = 588.6\text{N} \tag{7.20}$$

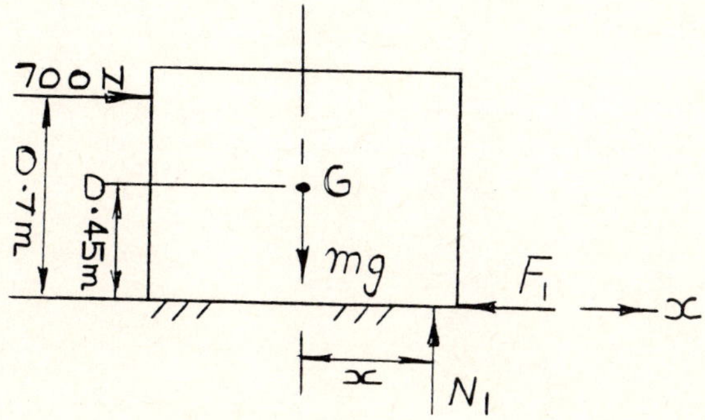

Figure 7.14

Applying Newton's 2nd Law in the 'x' direction, we get

$$700 - F_1 = m\ a \tag{7.21}$$

$$\text{but } F_1 = \mu_k\ N_1 \tag{7.22}$$

$$= 0.3 \times 588.6 = 176.58 \text{ N}$$

Substituting equation (7.22) into equation (7.21), we get

$$700 - 176.58 = 60 \times a$$

$$\therefore a = 8.724 \text{ m/s}^2$$

From equation (7.15)

$$\sum M_G = 0$$

$$\text{or } 700 \times 0.25 + F_1 \times 0.45 = N_1 \times x$$

$$\text{or } 175 + 176.6 \times 0.45 = 588.6\ x$$

$$\therefore x = 0.432 \text{ m}$$

As $x < 0.6$ m, the crate will slip and not overturn. If the crate overturned, a different calculation would have to be carried out.

Example 7.8

A beam of AB of mass 90 kg is supported by two equal lengths rods of negligible mass, as shown in Figure 7.15. If the angular velocity of the rods is 5 rad/s, clockwise, determine the acceleration of AB and the forces in the rods when $\theta = 25°$.

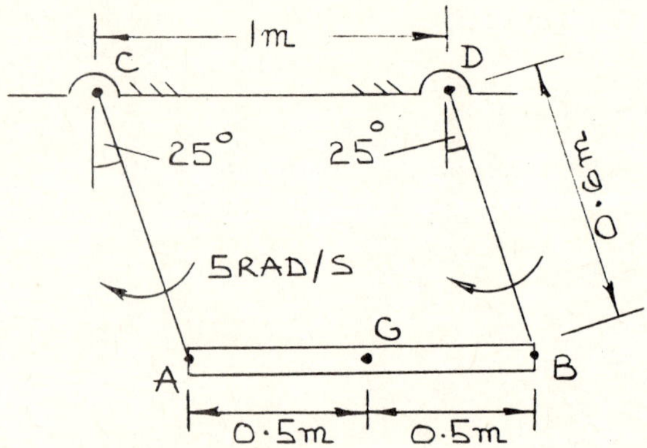

Figure 7.15

The free body diagram is shown in Figure 7.16, together with the 'n' and 't' directions.

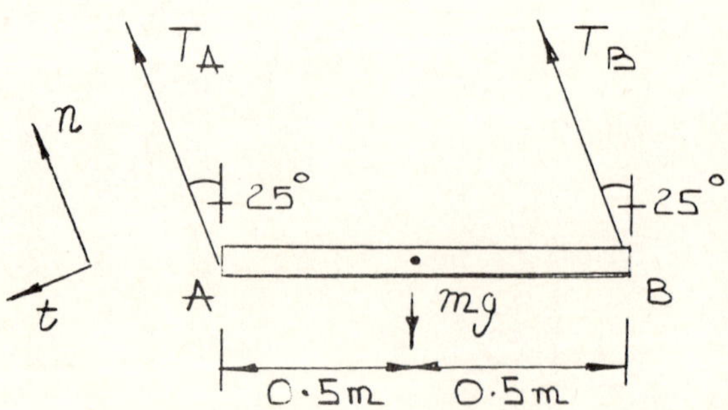

Figure 7.16

$$a_n = \omega^2 r = 5^2 \times 0.6 = 15 \text{ m/s}^2$$

Applying Newton's 2nd Law in the 'n' direction, we get

$$T_A + T_D - mg \cos 25° = ma_n$$

$$\text{or } T_A + T_D = 90 \times 15 + 90 \times 9.81 \cos 25° \tag{7.23}$$

$$= 1350 + 800.2$$

$$\text{or } T_A + T_D = 2150.2$$

Applying Newton's 2nd Law in the 't' direction, we get

$$mg \sin \theta = m\, a_t$$

$$\text{or } a_t = \frac{90 \times 9.81 \times \sin 25°}{90}$$

$$a_t = 4.146 \text{ m/s}^2$$

From equation (7.15)

$$\sum M_G = 0$$

$$\text{or } T_A \cos 25° \times 0.5 = T_D \cos 25° \times 0.5 \tag{7.24}$$

$$\text{or } T_A = T_D$$

Substituting equation (7.24) into equation (7.23),

$$T_A = T_D = 2150.2/2 = 1075.1 \text{ N} \tag{7.25}$$

$$\text{Now } a = \sqrt{\left(a_n^2 + a_t^2\right)} \tag{7.26}$$

$$= \sqrt{\left(15^2 + 4.146^2\right)} = 15.562 \text{ m/s}^2$$

Problems 7.2

1. A square plate is pinned to a block at one of its corners, as shown in Figure 7.17. If the block is subjected to a horizontal acceleration '*a*', determine the value of '*a*' which will keep the edge of the plate BC in a vertical position. Neglect frictional effects.

[$a = g$]

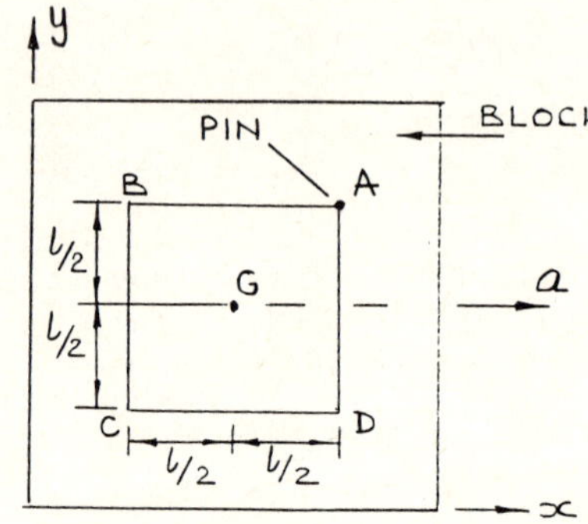

Figure 7.17

2. An 'L' shaped framework of uniform section is pinned to a block at the point 'A', as shown in Figure 7.18. Determine the horizontal acceleration '*a*' that will keep the side AB in a vertical position. Neglect friction.

[$a = g/3$ to the left]

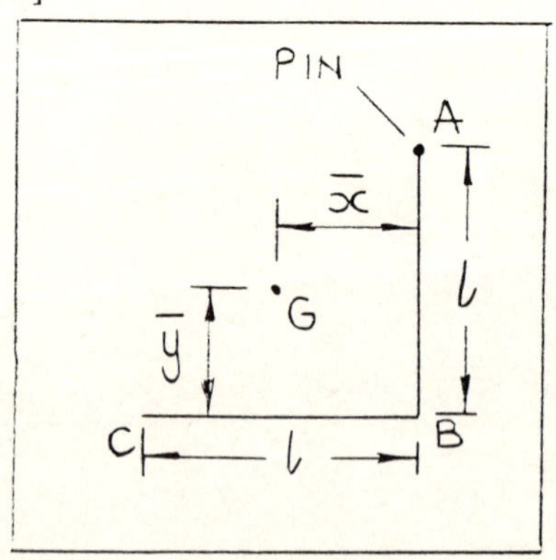

Figure 7.18

3. A crate rolls on small wheels on a horizontal surface, as shown in Figure 7.19. The crate is subjected to a horizontal force '*P*':

(a) If this force is applied to the top of the crate, determine its minimum value so that the crate will overturn at point ‘A’;

(b) If this force is applied to the bottom of the crate, determine its minimum value so that the crate will overturn at the point ‘B’.

[$mg_b h$; mgb/h]

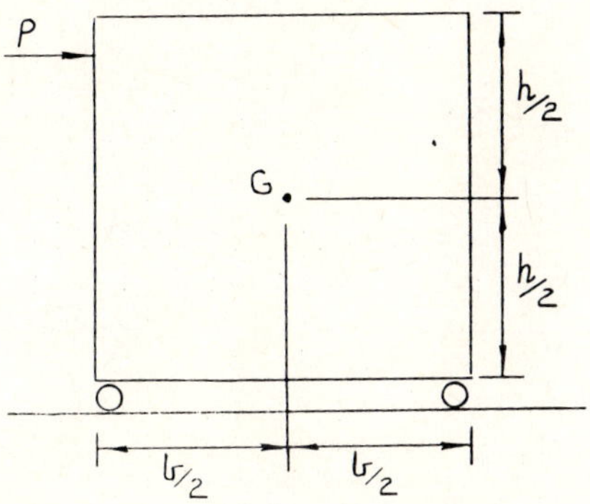

Figure 7.19

4. A horizontal beam AB of mass 50 kg is supported by two rods of negligible mass and of equal length, as shown in Figure 7.20. If the rods are subjected to a clockwise angular velocity of 4 rad/s, determine the acceleration of the beam and the forces in the rods, in the position shown.

[$a = 11.943$ m/s^2; $T_{AC} = T_{BD} = 502.3$ N]

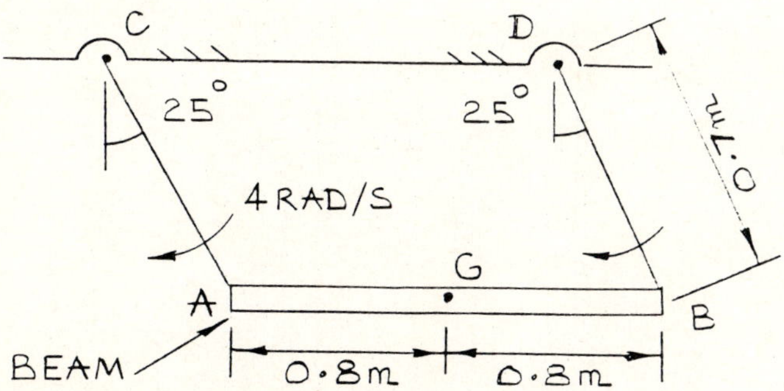

Figure 7.20

5. A uniform section pole ‘AB’ of mass 50 kg is hinged to the back of a truck at the point ‘A’ and it is kept in place by a wire ‘BC’ of negligible mass, as shown in Figure 7.21. If the truck accelerates on a horizontal surface with a value of acceleration of

4 m/s^2, in the direction shown, determine the total force that acts at the hinge 'A'.

[914.6 N]

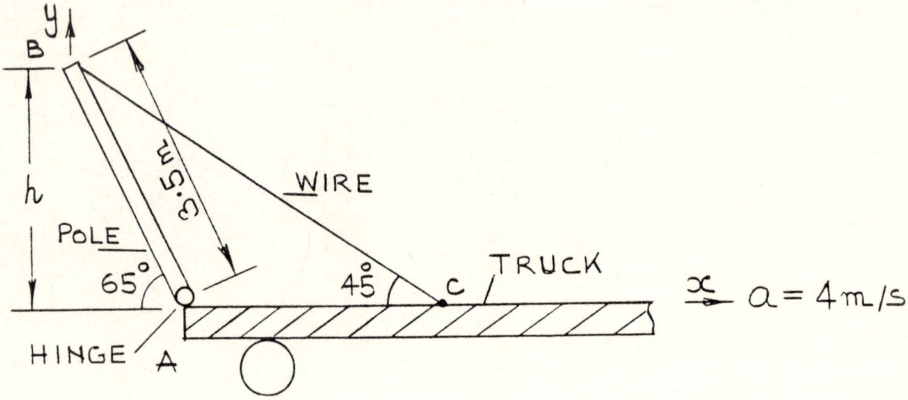

Figure 7.21

6. A horizontal uniform section beam 'AB' of mass 'm', is rigidly attached at its end 'A' to a parallelogram linkage, as shown inf Figure 7.22. The motion of the system is such that the beam always remains horizontal. If the link 'CD' is given a counter-clockwise angular velocity 'ω', determine expressions for the maximum shearing force and bending moment.

[$F_A = m\,(g + \omega^2\, r \cos\theta)$; $M_A = ml\,(g + \omega^2\, r\cos\theta)/2$]

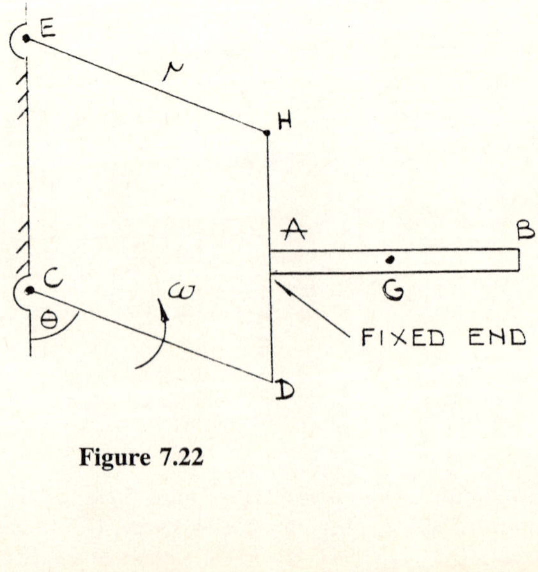

Figure 7.22

7.3.2 Rotation About a Fixed Axis 'O'

The equations of motion for general plane motion are:

$$\sum M_G = I_G \bar{\alpha}$$

$$\text{and} \sum F = m\bar{a}$$

where $\bar{a}$ consists of normal and tangential components, namely a_n and a_t, respectively, where

$$a_n = \omega^2 r_G$$

$$a_t = \alpha r_G$$

where r_G is the distance between the centre of gravity of the rigid body and the point about which it is rotating, as shown in Figure 7.23.

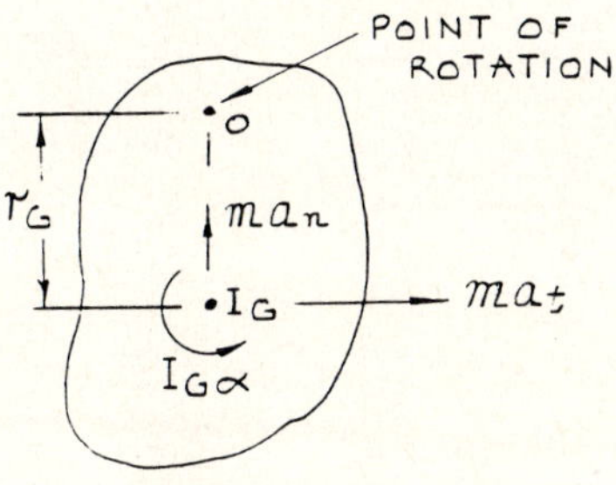

Figure 7.23

In Figure 7.23, the sum of the moments about 'O' is:

$$\sum M_o = \sum M_G + m\, a_t\, r_G$$

$$= I_G\, \alpha + m\, r_G^2\, \alpha$$

$$\text{or} \sum M_o = I_o\, \alpha \qquad (7.27)$$

where

I_o = the mass moment of inertia of the body about 'O'

$$= I_G + m\, .\, r_G^2$$

Example 7.9

A disk of uniform thickness and of a mass of 40 kg is allowed to rotate about the pin 'O'. If the disk is subjected to a tangential force of 25 N on its outer circumference, as shown in Figure 7.24, determine its angular acceleration and its bearing forces at 'O'. It may be assumed that the frictional resistance torque is 1.5 Nm.

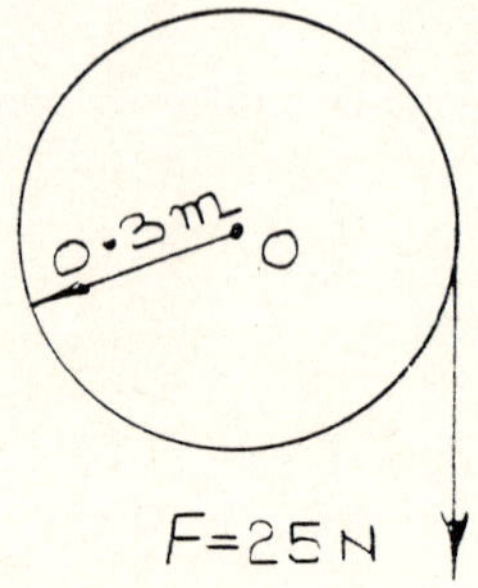

Figure 7.24

$$I_o = I_G = m\,r^2/2 = 40 \times 0.3^2/2 = 1.8 \text{ kg . m}^2$$

$$\sum M_o = I_o\,\alpha$$

$$\text{or } 25 \times 0.3 - 1.5 = 1.8\,\alpha$$

$$\text{or } \alpha = 3.333 \text{ rad/s}^2$$

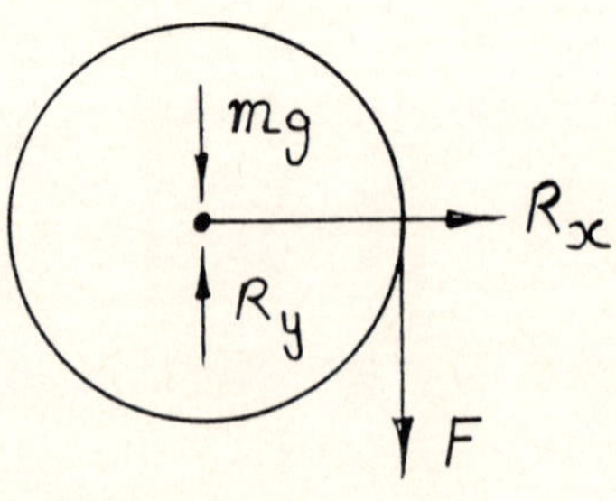

Figure 7.25 Free Body Diagram

$$\text{In } x \text{ direction, } R_x = m\,a_x = 0$$

$$\text{In } y \text{ direction, } F + mg - R_y = m\,a_y = 0$$

$$\therefore R_y = 25 + 40 \times 9.81$$

$$R_y = 417.4 \text{ N}$$

Example 7.10

A uniform thickness disc of mass 50 kg, has attached to it a mass of magnitude 15 kg, via a wire of negligible mass, as shown in Figure 7.26. The disc rotates about the bearing 'O', where the frictional torque is 1.2 Nm. Determine the angular acceleration of the disc and the tension in the wire.

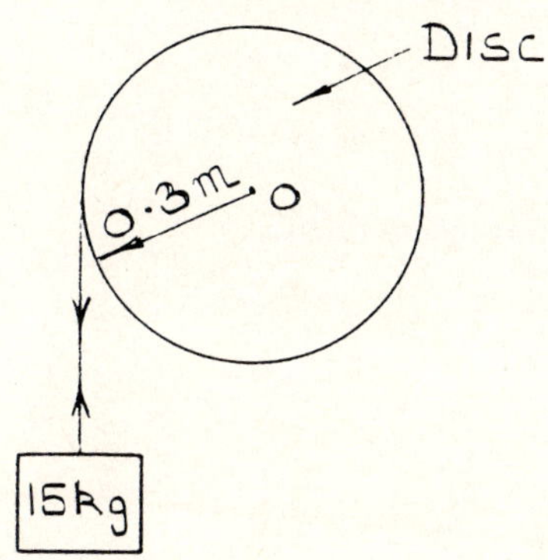

Figure 7.26

$$I_o = I_G = mr^2/2 = 50 \times 0.3^2/2$$

$$I_o = 2.25 \text{ kg . m}^2$$

Consider Motion of the Disc

$$M_o = I_o\,\alpha$$

$$\text{or } T \times 0.3 - 1.2 = 2.25\,\alpha \tag{7.28}$$

$$\text{or } T = (2.25\,\alpha + 1.2)/0.3$$

$$T = 7.5\,\alpha + 4$$

Consider Motion of the Mass

$$T - mg = -ma$$

$$\text{or } T = m(g - a)$$

$$\text{but } a = \alpha r \tag{7.29}$$

$$\therefore T = 15(9.81 - \alpha \times 0.3)$$

$$= 147.15 - 4.5\,\alpha$$

Equating (7.28) and (7.29), we get

$$7.5\,\alpha + 4 = -4.5\,\alpha + 147.15,$$

$$\text{or } 12\,\alpha = 143.15 \tag{7.30}$$

$$\alpha = 11.93 \text{ rad/s}^2$$

Substituting equation (7.30) into equation (7.28)

$$T = 7.5 \times 11.93 + 4$$

$$T = 93.47 \text{ N} \tag{7.31}$$

Example 7.11

A slender uniform section rod AB, of mass '*m*', is pinned at 'A'. If the rod is released from rest when $\theta = 0°$, determine the force in the pin when $\theta = 60°$.

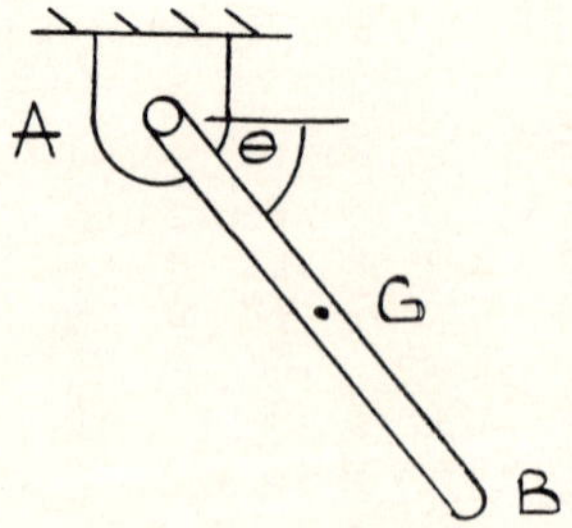

Figure 7.27

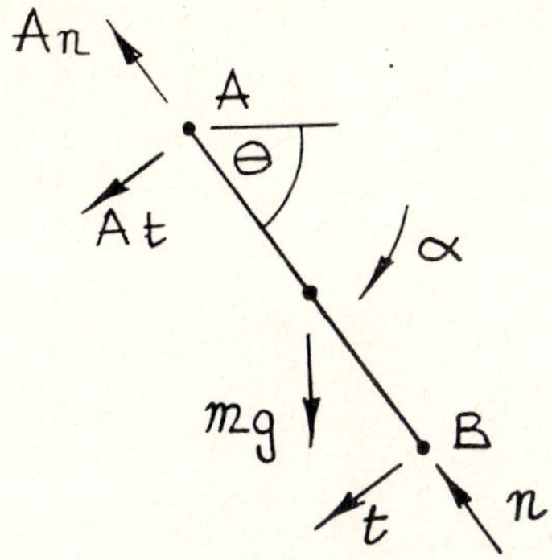

Figure 7.28 Free Body Diagram

$I_A = ml^2/3$

At the centre of gravity `G′

$a_n = \omega^2 \ell/2$

$a_t = \alpha \ell/2$

Applying Newton's 2nd Law in the '*n*' direction

$$A_n - mg \sin\theta = m a_n \tag{7.32}$$

$$= m\omega^2 \ell/2$$

Applying Newton's 2nd Law in the '*t*' direction

$$A_t + mg \cos\theta = m \alpha \ell/2 \tag{7.33}$$

From equation (7.27),

$$\sum M_A \quad mg \cos\theta \left(\ell/2\right) = I_A \alpha \tag{7.34}$$

$$\text{or } mg \cos\theta \left(\ell/2\right) = \left(m\ell^2/3\right) \alpha$$

$$\text{or } \alpha = \frac{g \cos\theta \left(\ell/2\right)}{\ell^2/3} = 1.5\, g \cos\theta/\ell \tag{7.35}$$

From equation (6.12)

$$\alpha \, d\theta = \omega \, d\omega$$

$$\text{or} \int_o^{\omega} \omega \, d\omega = (1.5 \; g/\ell) \int_o^{\omega} \cos\theta \, . \, d\,\theta, \tag{7.36}$$

$$\text{or } \omega^2/2 = (1.5 \; g/\ell) \sin \theta$$

When $\theta = 60°$

$$\omega^2 = (3 \; g/\ell) \times 0.866,$$

$$\text{or } \omega^2 = 25.486/\ell \tag{7.37}$$

$$\text{and } \omega = 5.049/\ell^{\frac{1}{2}}$$

$$\text{Now } \alpha = 12.744/\ell \tag{7.38}$$

Substituting equation (7.38) into equation (7.33)

$$A_n = m \times 25.486 \; (\ell/2)/\ell + m \times 9.81 \times 0.866$$

$$= 12.743 \; m + 8.496 \; m$$

$$A_n = 21.239 \; m$$

Substituting A_n into equation (7.33)

$$A_t = m \times 12.744 \; (\ell/2)/\ell - m \times 9.81 \times 0.5$$

$$= 6.372 \; m - 4.905 \; m$$

$$A_t = 1.467 \; m$$

$$A = \sqrt{(A_n^2 + A_t^2)}$$

$$= \sqrt{(21.239^2 + 1.467^2)} \; m$$

$$A = 21.29 \; m$$

Problems 7.3

1. A uniform thickness flywheel of mass 20 kg is free to rotate about the point 'O'. If the flywheel is subjected to a constant downward force 10 g N, what will be the angular acceleration of the flywheel, where 'g' = acceleration due to gravity? Neglect frictional effects.

[α = 49.05 rad/s^2]

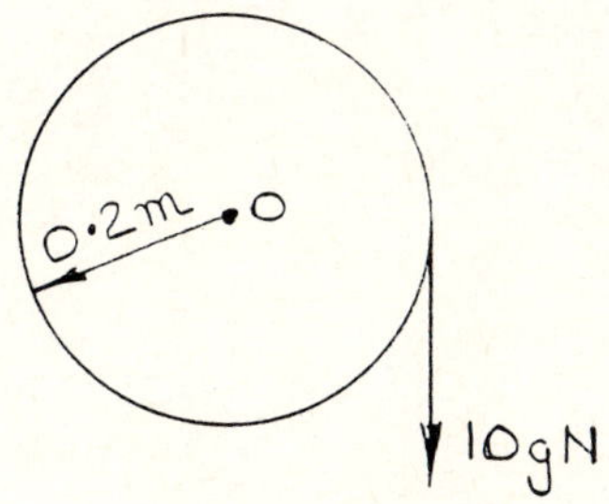

Figure 7.29

2. If instead of the 10 g N force, the above flywheel has a mass of 10 kg attached to it, via a light cord, what will be the angular acceleration of the flywheel and the tension in the cord?

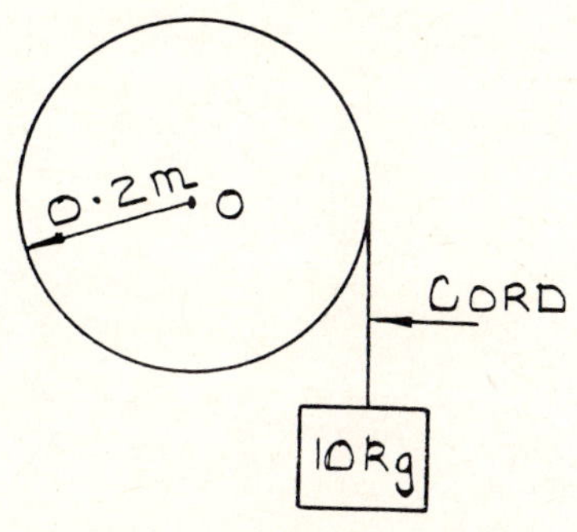

Figure 7.30

3. A pendulum of mass 20 kg and with a radius of gyration of 0.4 m about the point 'O', is free to rotate about this point. If the pendulum is released from rest, when θ = 0°, determine the value of the reaction at the pin, when θ = 65°. The centre of gravity of the pendulum is at the point 'G' shown in Figure 7.31.

[O = 205.9 N]

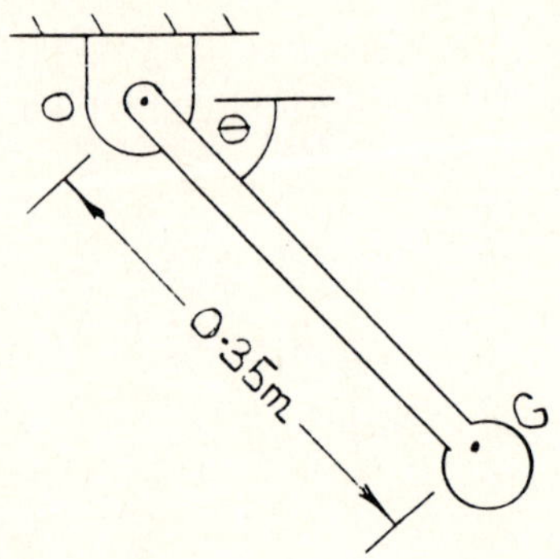

Figure 7.31

4. A single blade of a propeller is rotating with an angular velocity of 4 rad/s and an angular acceleration of 3 rad/s^2, as shown in Figure 7.32. The mass of the propeller blade is 1.5 kg and its radius of gyration about its centre of gravity 'G' is 0.1 m. If at a certain instant, the propeller blade is vertically upwards, determine the bending moment shearing force and normal reaction at the hub 'A'.

[M = 0.383 Nm; F = 1.35 N; R = 7.515 N]

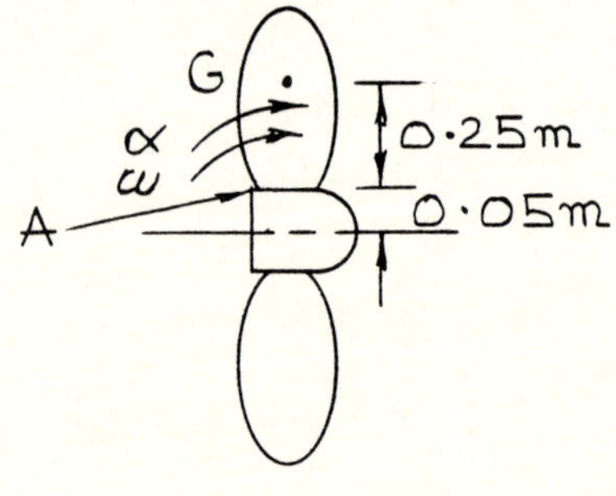

Figure 7.32

5. A thin-curved hemispherical beam, rather like the shape of a garden sickle, but of uniform cross-section, is subjected to a counter-clockwise angular velocity of 6 rad/s and a counter-clockwise angular acceleration of 5 rad/s^2, both acting in a horizontal plane, as shown in Figure 7.33. If the mass of the curved beam is 0.5 kg and its mean radius is 0.1 m, determine the bending moment and shearing force that occur at the point 'A' due to this motion.

[M_A = 0.0445 N m; F_A = 1.385 N]

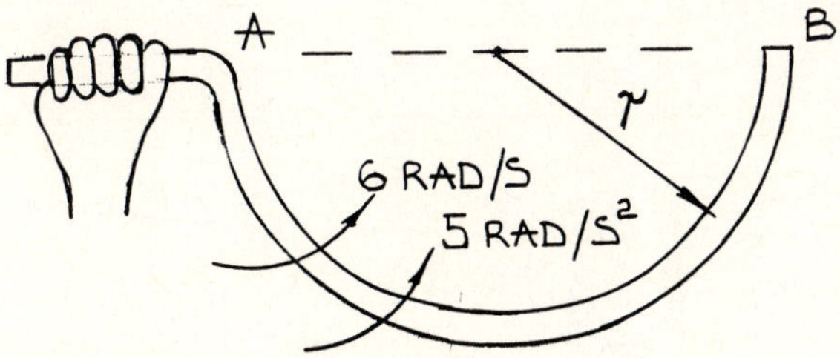

Figure 7.33

6. A pencil which is initially held in a perpendicular position to a flat horizontal surface, is suddenly released. If the pencil slips at 60° to the horizontal, determine the coefficient of friction between the pencil and the horizontal surface.

[μ_s = 0.188]

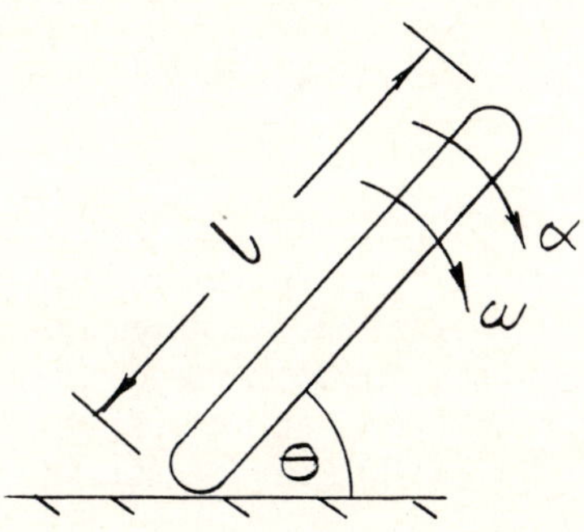

Figure 7.34

7.3.3 General Plane Motion

In this case, the equations of motion are:

$$\sum F_x = m\overline{a}_x \tag{7.39}$$

$$\sum F_y = m\,\overline{a}_y \tag{7.40}$$

$$\sum M_G = I_G\,\overline{\alpha} \tag{7.41}$$

Sometimes, however, instead of equation (7.41), it may be convenient to use the following equation of motion:

$$\sum M_p = I_G \alpha + m \bar{a} d \qquad (7.42)$$

where $\bar{a}$ and d are shown in Figure 7.35(b) and ΣM_P = the sum of the moments about the point 'P'

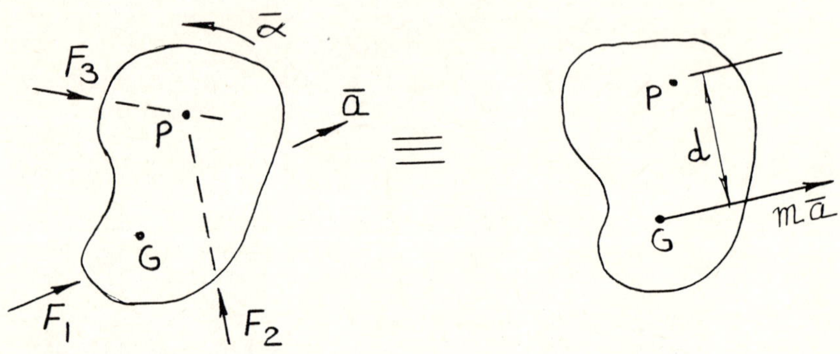

(a) Free Body Diagram **(b) Kinetic Diagram**

Figure 7.35

Example 7.12

A spool which is initially at rest has a light cord wrapped around its hub, as shown in Figure 7.36. If the cord on the left side of the hub is subjected to a vertical force of 70 N, which causes the spool to move upwards, determine the angular velocity of the spool after 2.5 seconds. Determine also the tension 'T' in the cord shown in Figure 7.36. It may be assumed that the mass of the spool is 7 kg and its radius of gyration is 0.25 m. It may also assumed that there is no slipping.

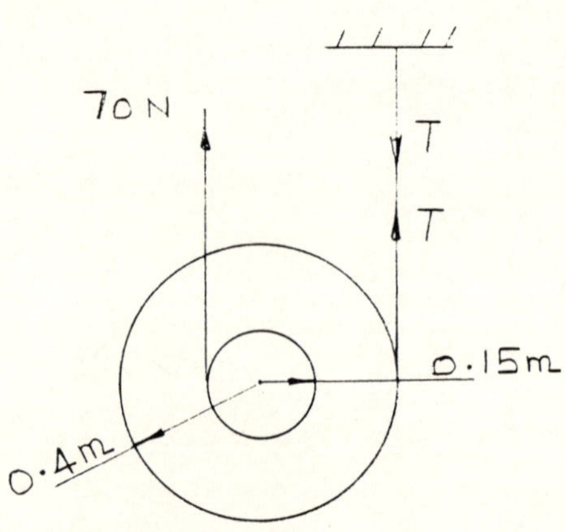

Figure 7.36

The free body diagram is shown in Figure 7.37.

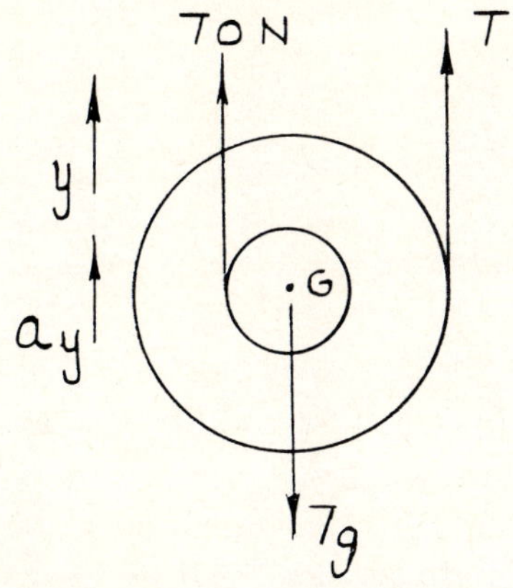

Figure 7.37 Free Body Diagram

$$I_G = mk^2 = 7 \times 0.2^2 = 0.28 \text{ kg . m}^2$$

Applying the Newton's 2nd Law in the 'y' direction, we get

$$70 + T - 7g = 7\,a_y$$
$$\text{or } T = 7\,a_y - 1.33 \qquad (7.43)$$

$$\text{Now} \sum M_G = I\,\alpha$$
$$\text{ie } 70 \times 0.15 - T \times 0.4 = 0.28\,\alpha \qquad (7.44)$$

$$\text{but } a_y = \alpha \times 0.4$$
$$\text{or } \alpha = 2.5\,a_y \qquad (7.45)$$

Substituting equation (7.45) into equation (7.44)

$$10.5 - 0.4\,T = 0.28 \times 2.5\,a_y = 0.7\,a_y$$
$$\text{or } T = 26.25 - 1.75\,a_y \qquad (7.46)$$

Equating (7.43) and (7.46)

$$7\,a_y - 1.33 = 26.25 - 1.75\,a_y$$

$$\text{or } 8.75\,a_y = 27.58 \qquad (7.47)$$

$$\text{or } a_y = 3.152 \text{ m/s}^2$$

$$\text{and } \alpha = 7.88 \text{ rad/s}^2$$

Substituting equation (7.47) into equation (7.43), we get

$$T = 12.46 \text{ N}$$

From equation (6.14)

$$\omega = \omega_o + \alpha t$$

$$= 0 + 7.88 \times 2.5$$

$$\omega = 19.7 \text{ rad/s}$$

Example 7.13

A wheel of mass 20 kg is subjected to a couple of 40 Nm as shown in Figure 7.38. If the radius of gyration of the wheel is 0.2 m and the coefficients of static and kinematic friction are 0.32 and 0.24 respectively, determine the acceleration of the centre of gravity of the wheel.

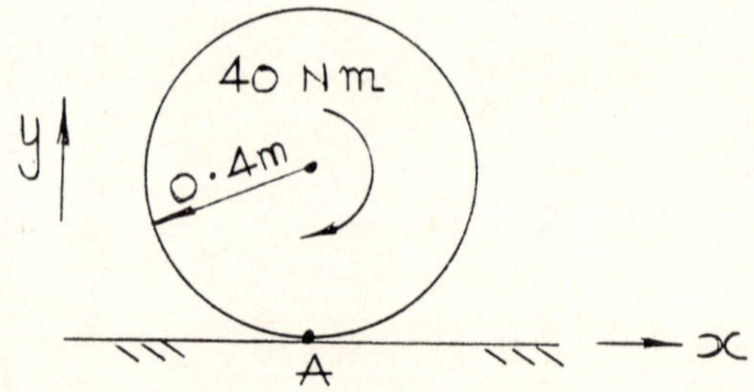

Figure 7.38

$$I_G = mk^2 = 20 \times 0.2^2 = 0.8 \text{ kg . m}^2$$

The free body diagram is shown in Figure 7.39.

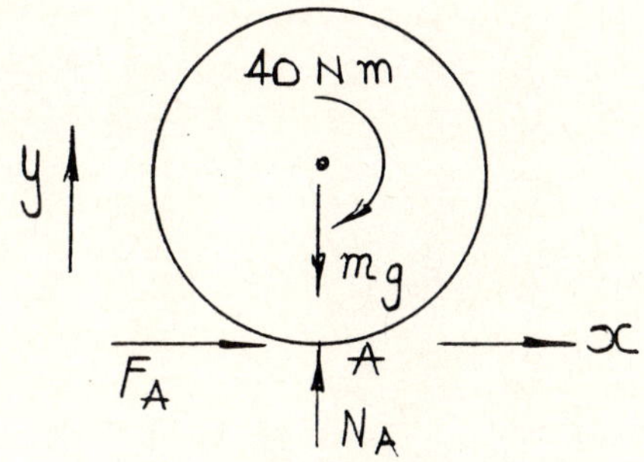

Figure 7.39 Free Body Diagram

In '*x*' Direction

$$F_A = m\,a_x = 20\,a_x,$$
$$\text{or } a_x = F_A/20 \qquad (7.48)$$

In '*y*' Direction

$$N_A = mg = 192.6 \text{ N} \qquad (7.49)$$

$$\text{Now } \sum M_G = I_G\,\alpha$$
$$40 - F_A \times 0.4 = 0.8\,\alpha \qquad (7.50)$$

Now if **no slipping takes place**

$$a_x = \alpha \times 0.4$$
$$\text{or } \alpha = a_x/0.4 \qquad (7.51)$$

or $40 - 0.4\ F_A = 0.8\ a_x/0.4$ (7.52)

or $40 - 0.4\ F_A = 2\ a_x$ (7.53)

Substituting equation (7.48) into equation (7.53)

$$40 - 0.4\ F_A = 2 \text{ x } F_A/20$$

$$\text{or } F_A = 80 \text{ N}$$

To check for slipping

$$\mu_s\ N_A = F_A$$

$$\text{or } \mu_s = F_A\ /\ N_A = 80/196.2$$

$$\mu_s = 0.4$$

ie as $\mu_s > 0.32$, the **wheel will slip**

$$\therefore\ F_A = \mu_k\ N_A = 0.24 \text{ x } 196.2$$

$$\text{or } F_A = 47.1 \text{ N} \qquad (7.54)$$

Substituting equation (7.54) into equation (7.48)

$$a_x = 47.1\ /\ 20 = 2.355 \text{ m/s}^2$$

Example 7.14

A rod of mass 20 kg has connected to its ends two small wheels of negligible mass. If the rod is held in the position shown and it is suddenly released, determine the angular acceleration of the rod and the reactions at ‘A’ and ‘B’.

The motion is constrained so the accelerations at ‘A’ and ‘B’ will be as shown in Figure 7.41. Additionally, the acceleration diagram will be as shown in Figure 7.42.

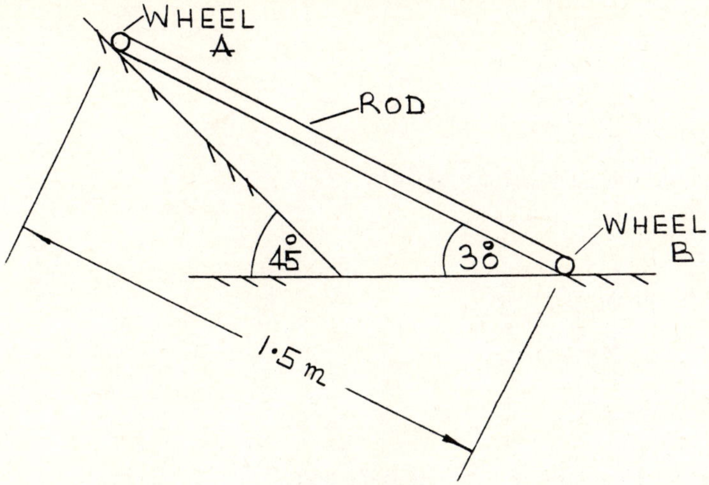

Figure 7.40

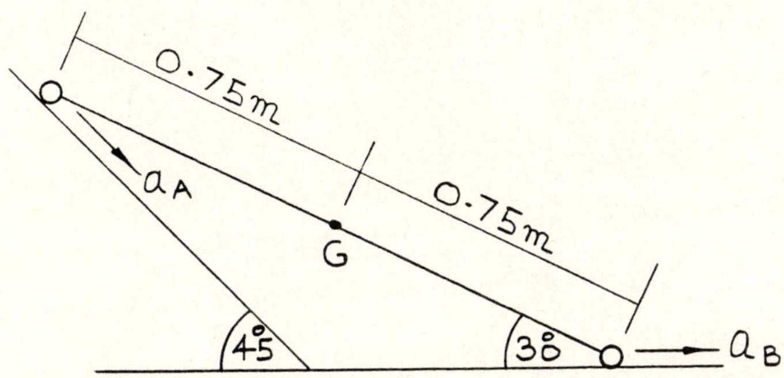

Figure 7.41

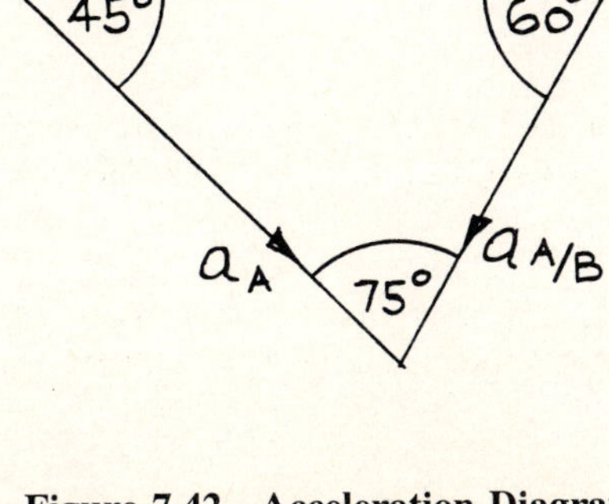

Figure 7.42 Acceleration Diagram

Now $a_A = a_B + a_{A/B}$

and $a_{A/B} = 1.5\alpha$

Hence, from the sine rule of equation (1.6), we get

$$\frac{1.5\alpha}{\sin 45°} = \frac{a_A}{\sin 60°} = \frac{a_B}{\sin 75°}$$

$\therefore\ a_A = 1.837\ \alpha$

and $a_B = 2.049\ \alpha$

From the vector diagram of Figure (7.43) the acceleration at 'G' is given by

$$a_G = a_B + a_{G/B}$$

$$= 2.049\ \alpha + 0.75\ \alpha$$

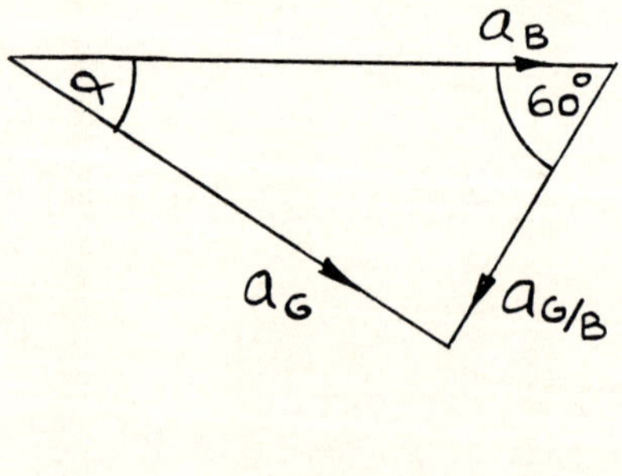

Figure 7.43

Now from Figure 7.43

$(a_G)_x = 2.049\ \alpha - 0.75\ \alpha \cos 60° = 1.674\ \alpha$

and $(a_G)_y = -0.75\ \alpha \sin 60° = -0.6495\ \alpha$

Consider now the kinetics of motion of the rod, as shown in Figure 7.44.

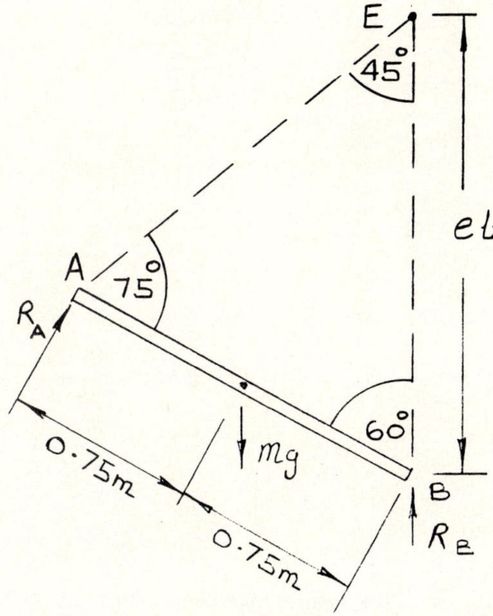

Figure 7.44

Applying the sine rule to Figure 7.44

$$\frac{1.5}{\sin 45} = \frac{eb}{\sin 75}$$

$$\therefore\ eb = 2.049 \text{ m}$$

$$\text{Now } I_G = m\,\ell^2/12 = 20 \times 1.5^2/12 = 3.75 \text{ kg . m}^2$$

$$\sum M_E = mg \times 0.75 \cos 30° = m\left(a_G\right)_x \times \left(2.049 - 0.75 \sin 30°\right)$$

$$+\ m\left(a_G\right)_y \times 0.75 \cos 30° + I_G\,\alpha$$

$$\text{or } 127.4 = 31.48\,\alpha \times 1.674 + 8.437\,\alpha + 3.75\,\alpha$$

$$\text{or } 127.4 = \left(52.698 + 8.437 + 3.75\right)\alpha$$

$$\therefore\ \alpha = 1.96 \text{ rad/s}^2$$

In 'x' Direction

$$R_A \sin 45° = m\,(a_G)_x = 20 \text{ x } 1.674 \text{ x } 1.96$$

$$= 65.62$$

$$\text{or } R_A = 92.81 \text{ N}$$

In 'y' Direction

$$R_B + R_A \cos 45 - mg = m\,a_y$$

$$R_B + 0.707 \text{ x } 92.81 - 196.2 = -20 \text{ x } 0.6495 \text{ x}1.96$$

$$\text{or } R_B = 105.1 \text{ N}$$

Problems 7.4

1. A wheel of mass 20 kg and with a centroidal radius of gyration of 0.1 m, rolls down a slope of 25° without slipping. Determine the frictional force between the wheel and the surface, and the minimum coefficient of static friction to prevent slipping.

   [F = 16.59 N; μ_s = 0.093]

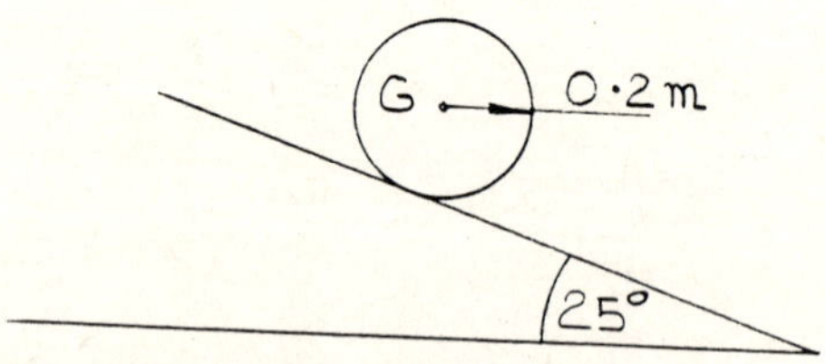

Figure 7.45

2. A thin-walled ring of mass 2 kg and of radius 0.5 m rolls down a slope of 25° to the horizontal. If the coefficients of static friction and kinetic friction between the ring and the surface are 0.3 and 0.2 respectively, determine the angular acceleration and the distance travelled by the ring in 2.5 seconds.

   [α = 3.556 rad/s^2; S = 7.4 m]

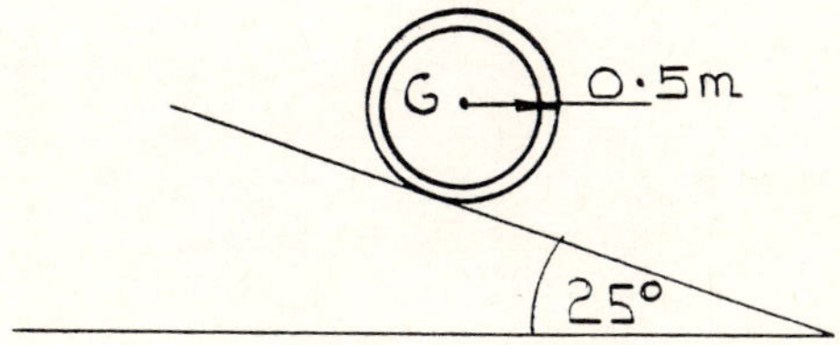

Figure 7.46

3. A horizontal uniform section beam is hinged firmly at the point 'A' and is supported by a light cord at 'B'. If the cord suddenly breaks determine:

(i) the acceleration at the point 'B';
(ii) the reaction at the point 'A'.

[$a_B = g; \quad R_A = 3\,mg/4$]

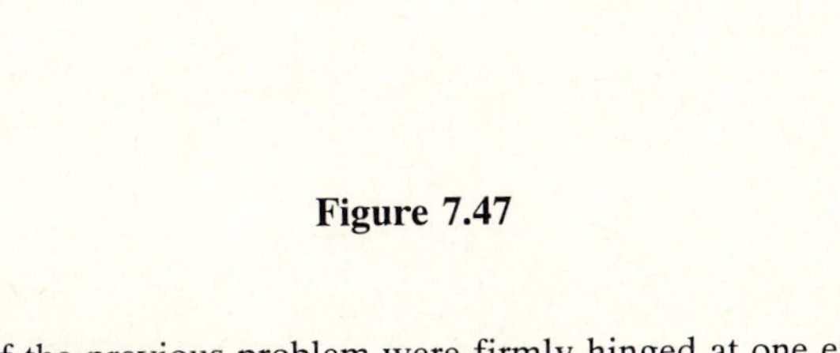

Figure 7.47

4. If the beam of the previous problem were firmly hinged at one end and supported by a light cord at B, determine:

(i) the acceleration at 'B';
(ii) the reaction at 'A' when the cord breaks.

[$a_B = 3\,g/2; \quad R_A = mg/4$]

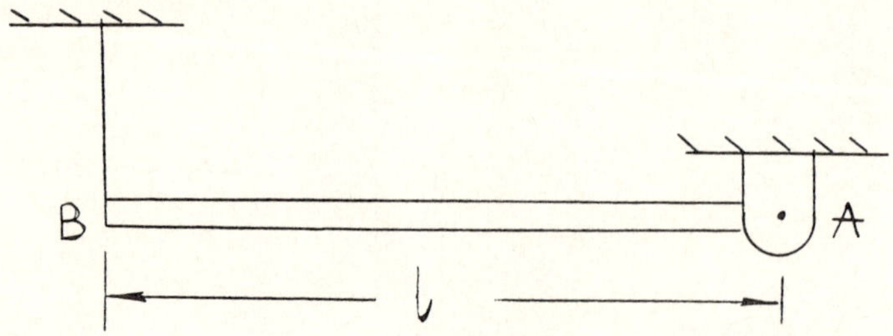

Figure 7.48

5. A crank AB rotates about 'A' with counter-clockwise angular velocity of 3 rad/s and a counter-clockwise angular acceleration of 5 rad/s^2, and drives the mechanism shown in Figure 7. If the bar BC is of mass 1 kg and it remains in the horizontal position shown, determine the force in the rod CD. The rod CD is of negligible mass.

[F_{CD} = 5.132 N (tension)]

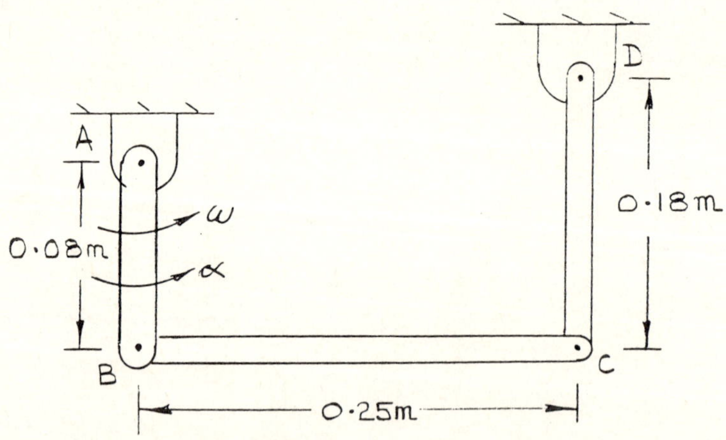

Figure 7.49

6. A rod of length 3 m and of mass 10 kg rests against a vertical wall at one end, while the other end B rests on a horizontal floor, on a small wheel, as shown in Figure 7.50. If the coefficient of kinetic friction between the rod and the vertical wall is 0.25, determine the acceleration at 'A', when the rod is suddenly released.

[a_A = 5.95 m/s^2]

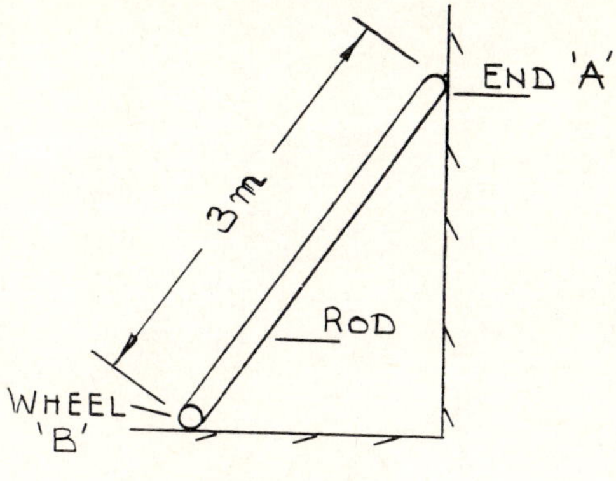

Figure 7.50

7.4 WORK, POWER AND ENERGY

In Chapter 4, we considered work, power and energy applied to a particle. Now for the translatory motion of a rigid body, the equations derived for a particle also apply, but as a rigid body can rotate, additional terms have to be considered. Consider a rigid body subjected to the couple shown in Figure 7.51(a) which causes rotation, as shown in Figure 7.51(b).

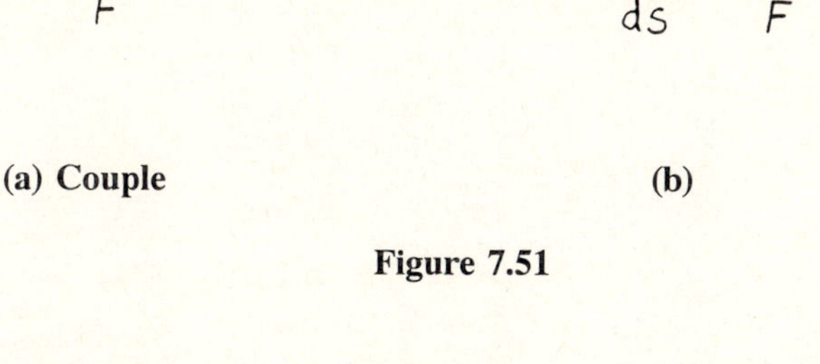

(a) Couple (b)

Figure 7.51

$$\therefore \text{ work done by a couple } = WD = \int M \, . \, d\theta \tag{7.56}$$

The work done by the couple $= \int 2F \,.\, \mathrm{d}s$

$$= \int 2F \,.\, \frac{b}{2} \,.\, \mathrm{d}\theta$$

$$= \int Fb \,.\, \mathrm{d}\theta$$

but $Fb = M =$ the value of the moment caused by the couple

Now kinetic energy $=$ KE $= mv^2/2$

but for a particle of a mass m_i, rotating on the rigid body at a distance r_i from the point 'O', as shown in Figure 7.52.

$$\text{KE} = \sum \frac{1}{2} m\, v^2 = \sum \frac{1}{2} m_i \,.\, r_i^2\, \omega^2$$

$$= \frac{1}{2} \omega^2 \sum m_i\, r_i^2 \qquad (7.57)$$

$$= \frac{1}{2} I_o\, \omega^2$$

where

$I_o = \Sigma\, m_i\, r_i^2$

= mass moment of inertia of the rigid body about 'O'

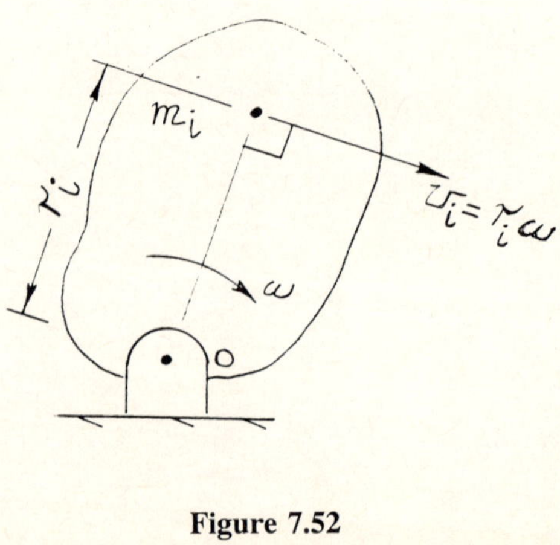

Figure 7.52

For general plane motion,

KE $= \Sigma \frac{1}{2} m_i v_i^2$ - see Figure 7.53

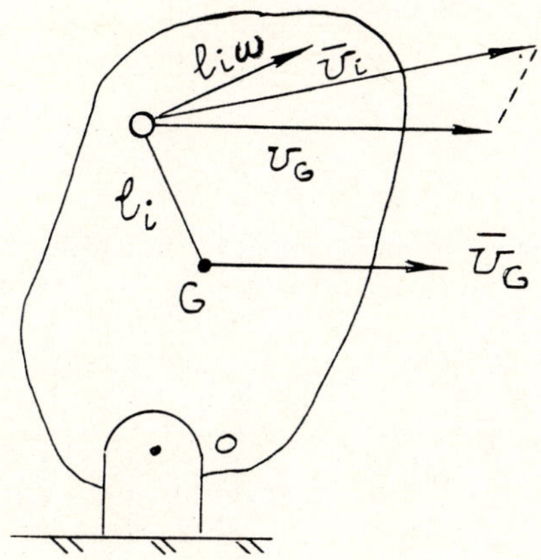

Figure 7.53

Applying the cosine rule of equation (1.7) to Figure 7.53

$$v_i^2 = \left(\overline{v_G^2} + \rho_i^2\,\omega^2 + 2\overline{v_G}\,\rho_i\,\omega\cos\theta\right)$$

Substituting equation (7.58) into the above equation for KE, we get

$$KE = \frac{1}{2}\sum m_i\left(\overline{v_G^2} + \rho_i^2\,\omega^2 + 2\overline{v_G}\,\rho_i\,\omega\cos\theta\right)$$

but $\omega\,\bar{v}\sum m_i\,\rho_i\cos\theta = \omega\,\overline{v_G}\sum m_i\,y_i = 0$

$$\therefore KE = \frac{1}{2}\sum m_i\left(\overline{v_G^2} + \rho_i\,\omega^2\right)$$

$$= \frac{1}{2} m\,\overline{v_G^2} + \frac{1}{2} I_G\,\omega^2 \qquad (7.59)$$

where

$\overline{v}_G$ = velocity of centre of gravity

I_G = mass moment of inertia of the body about its centre of gravity

By a similar process, it can be shown that

$$KE = \frac{1}{2} I_C \, \omega^2 \tag{7.60}$$

where I_C = mass moment of inertia about the instantaneous centre of zero velocity.

$$\text{Power} = P = \overline{F} \,.\, \overline{v}$$

$$\text{or } P = \frac{dU}{dt} = \overline{M_G} \frac{d\theta}{dt}$$

$$= \overline{M_G} \, \omega$$

where U = energy

$$\text{Total Power} = \overline{F} \,.\, \overline{v} + \overline{M_G} \, \omega \tag{7.61}$$

$$\text{KE} = \frac{1}{2} m \, \overline{v_G} \,.\, \overline{v_G} + \frac{1}{2} I_G \, \omega^2$$

$$\frac{d\,(\text{KE})}{dt} = \frac{1}{2} m \left(\overline{a_G} \,.\, \overline{v_G} + \overline{v_G} \,.\, \overline{a_G}\right) + I_G \,.\, \omega \,.\, \omega$$

$$= \overline{R} \,.\, \overline{v_G} + \overline{M_G} \, \omega$$

where

$\overline{R}$ = the resultant of all the forces acting on the body
$\overline{M_G}$ = the resultant of all the moments about the centre of gravity

Example 7.15

A block of mass 100 kg is attached to a light cord, which is wrapped around a drum, as shown in Figure 7.54. The drum is an integral part of a fly wheel and the combined mass moment of inertia of the drum and the flywheel is 20 kg . m^2. The frictional resistance couple exerted by the bearing is 40 Nm. If at a certain instant, the downward velocity of the block is 2 m/s, determine the velocity of the block after it falls downwards a further 1.5 m.

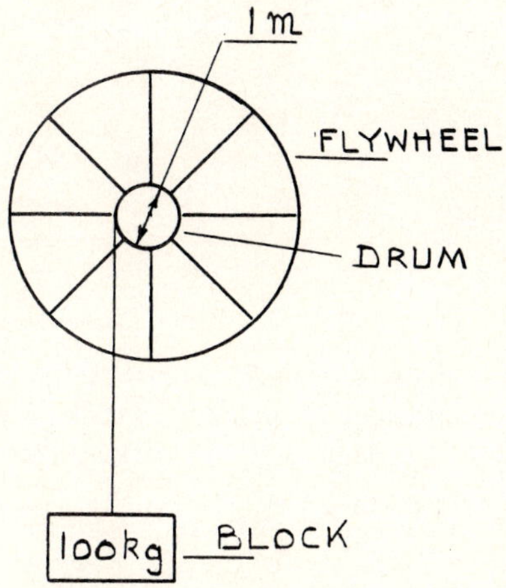

Figure 7.54

$$v_1 = 2 \text{ m/s} = \omega_1 r = \omega_1 \times 0.5$$

$$\therefore \omega_1 = 4 \text{ rad/s}$$

$$\text{Initial KE} = \frac{1}{2} m v_1^2 + \frac{1}{2} I \omega_1^2$$

$$= \frac{1}{2} \times 100 \times 2^2 + \frac{1}{2} \times 20 \times 4^2$$

$$= 200 + 160$$

$$= 360 \text{ J}$$

Potential energy = PE = mgh

or PE = 100 x 9.81 x 1.5 = 1471.5 J

Due to the block falling 1.5 m, the drum rotates by θ, where

$$\theta = \frac{1.5 \text{ m}}{0.5 \text{ m}} = 3 \text{ rad}$$

Now initial KE + PE = Final KE + Frictional work done, or

$$360 + 1471.5 = \frac{1}{2} \times 100 \times v_2^2 + \frac{1}{2} \times 20 \times \left(\frac{v_2}{0.5}\right)^2 + 40 \times 3$$

$$\text{or } 1711.5 = 50\ v_2^2 + 40\ v_2^2$$

$$\therefore\ v_2 = 4.361 \text{ m/s}$$

Example 7.16

A circular cylinder, a hoop and a sphere, each of mass '*M*' are each released from rest down a plane with a slope of 'θ'. Determine the velocity of each after their centres of gravity have fallen vertically a distance '*h*'. Assume that no slipping takes place.

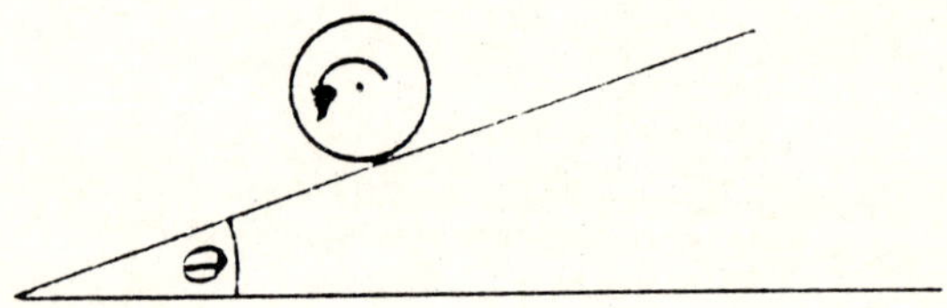

Figure 7.55

$$PE = mgh$$

$$\text{Final KE} = \frac{1}{2} mv^2 + \frac{1}{2} I \left(\frac{v}{r}\right)^2$$

$$= \frac{1}{2}\left(m + \frac{I}{r^2}\right) v^2$$

$$\text{PE} = \text{Final KE}$$

$$\text{or } mgh = \frac{1}{2}\left(m + \frac{I}{r^2}\right) v^2$$

$$\therefore\ v^2 = \frac{2\ mgh}{\left(m + I/r^2\right)}$$

For a solid circular cyliner $I = mr^2/2$

$$\therefore\ v^2 = \frac{2\ mgh}{\left[m + mr^2/(2r^2)\right]} = \frac{2\ gh}{\left(\frac{3}{2}\right)}$$

$$\text{and } v = 0.816\sqrt{2gh}$$

For a <u>hoop</u> $I = mr^2$

$$\therefore\ v^2 = \frac{2\,mgh}{(m + m)}$$

$$\text{and } v = 0.707\sqrt{2gh}$$

For a sphere, $I = 2mr^2/5$

$$\therefore\ v^2 = \frac{2\ mgh}{(m + 2\ m/5)}$$

$$\text{and } v = 0.845\sqrt{2\ gh}$$

Example 7.17

A slender beam of mass 5 kg is held in a horizontal position as shown in Figure 7.56, where the spring at 'B' has been initially compressed by 3 cm. If the spring is suddenly released, determine the angular velocity of the rod as it pass through its vertical position. The stiffness of the spring is 230,000 N/m and it is not attached to the beam. Determine also the reaction at the pin 'C' when the beam is in its vertical position.

$$\text{Strain energy} = SE = \frac{1}{2}k\ x^2$$

$$= \frac{1}{2} \times 230{,}000 \times (3 \times 10^{-2})^2$$

$$SE = 103.5\ \text{J}$$

$$\text{Final PE} = mgh = 5 \times 9.81 \times 0.8 = 39.24\ \text{J}$$

$$I_G = ml^2/12 = 5 \times 2^2/12 = 1.667\ \text{kg.m}^2$$

$$v_G = \omega \times 0.8$$

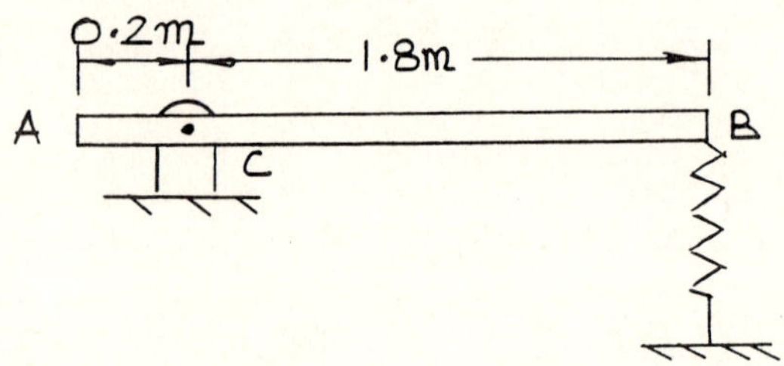

Figure 7.56

$$\text{Final KE} = \frac{1}{2}mv_G^2 + \frac{1}{2}I_G\,\omega^2$$

$$= \frac{1}{2} \times 5 \times (0.8\,\omega)^2 + \frac{1}{2} \times 1.667\,\omega^2$$

$$= 1.6\omega^2 + 0.833\,\omega^2$$

$$\text{Final KE} = 2.433\,\omega^2$$

$$\text{Now SE} = \text{Final PE} + \text{Final KE}$$

$$\text{or } 103.5 = 39.24 + 2.433\,\omega^2$$

$$\omega^2 = 26.41$$

$$\omega = 5.139 \text{ rad/s}$$

To calculate the reaction at pin 'C', see the free body diagram of Figure 7.57(a) and the kinetic diagram of Figure 7.57(b).

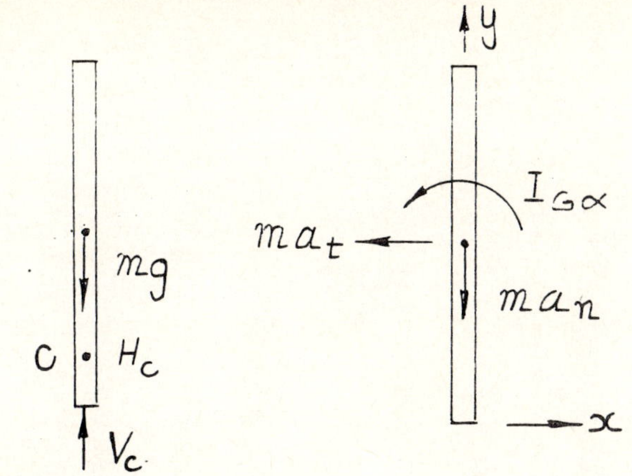

(a) Free Body Diagram **(b) Kinetic Diagram**

Figure 7.57

In x direction $H_C = ma_t = m\,r\,\alpha = 0$

In y direction

$$V_c - mg = -ma_n$$

$$\text{or } V_C = m\,(g - a_n)$$

$$= 5\,(9.81 - 21.128)$$

$$V_c = -56.59 \text{ N}\downarrow$$

Problems 7.5

1. A rotor of a generator was rotating with a speed of 2,000 rpm, when the generator was suddenly stopped. If the mass of rotor were 100 kg and its radius of gyration was 0.5 m, determine the number of additional revolutions that would have taken place before the rotor came to rest. It may be assumed that the frictional torque exerted by the rotor's bearings was 2.5 Nm.

   [θ = 34,893 revs]

2. A log is to be used as a battering ram, by suspending it on two light cables and deflecting the log through an angle 'θ', as shown in Figure 7.58. Determine the angle 'θ' that is required to cause the log to strike the object it is to batter, with a velocity of 7 m/s.

   [θ = 67.9°]

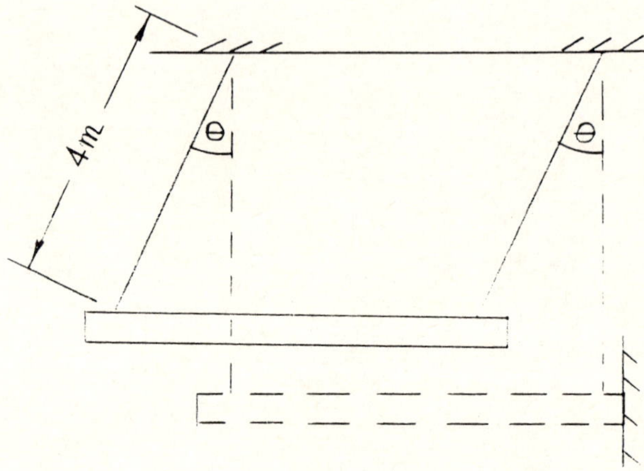

Figure 7.58

3. A wheel has two symmetrical outer hubs, on which it can roll, without slipping, on the inclined surface shown in Figure 7.59. If the wheel is subjected to a force of 80 N acting parallel to and up the plane, determine the angular velocity of the wheel after it travels up the plane a distance of 2 m, starting from rest. The mass of the wheel is 60 kg and its radius of gyration is 0.1 m. What is the power produced by this 80 N force when this value of 'ω' is reached?

[ω = 14.36 rad/s; P = 172.3 W]

4. A flywheel of mass 1000 kg and of radius of gyration 0.3 m is rotating at a speed of 4000 rpm when its speed is reduced to 200 rpm in 1½ minutes. Determine the average power exerted by the flywheel during this time.

[P = 88.26 kW]

5. Two slender rods, each of mass 1 kg, form the mechanism shown in Figure 7.59, where the point 'A' is a fixed axis of rotation and the point 'D' is on a light frictionless wheel. If the mechanism is suddenly released from rest, when the angle θ = 65°, determine the velocity of the point 'D' when θ = 30°.

[v_D = 2.59 m/s]

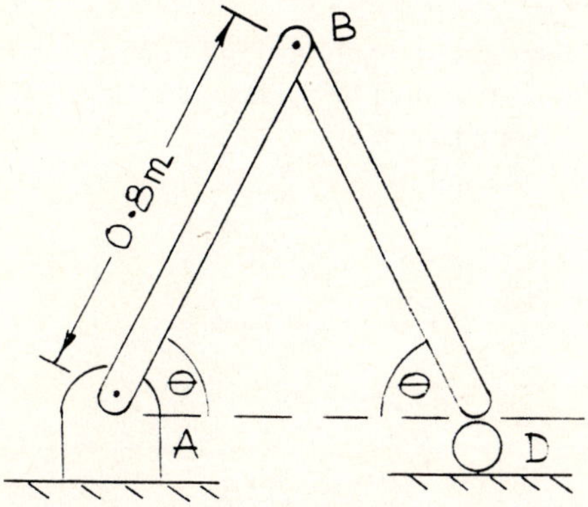

Figure 7.59

6. A mechanism consists of two rods AB and BD pinned together at 'B' and with a pin fitted through the smooth collar fitted at 'D'. The collar fits snugly over a vertical rod, so that its motion is constrained in the vertical direction. If the mechanism is released from the position shown in Figure 7.60, determine the velocity of 'B' when the rod AB moves through 90°. The mass of AB = 2 kg, the mass of BD = 5 kg and the mass of the collar = 3 kg.

[v_B = 2.11 m/s]

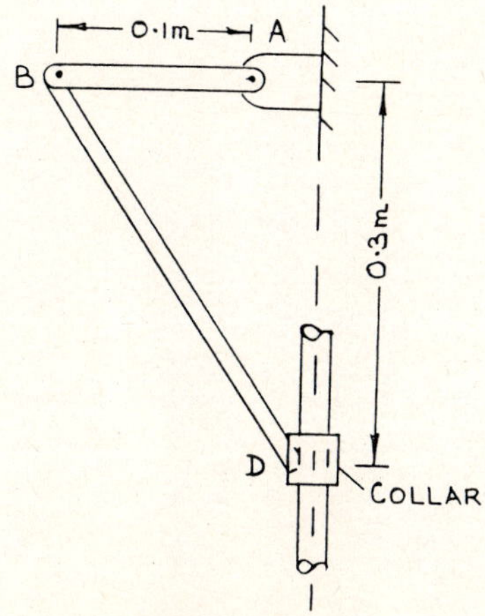

Figure 7.60

7.5 IMPULSE AND MOMENTUM

In the case of a rigid body, we have to take into account, the angular impulse and momentum in addition to the linear impulse and momentum.

7.5.1 Linear Impulse and Momentum

The equations for particle motion also apply to rigid body motion, for problems involving linear impulse and momentum. If a force 'F' were applied to the centre of gravity of a body of mass '*m*', the following equations apply:

$$\int F \,.\, \mathrm{d}t = m\left(v_G\right) - m\left(v_G\right)_1 , \qquad (7.62)$$

where

$\int F \,.\, \mathrm{d}t$ = linear impulse

$\left(v_G\right)_2$ = final velocity

$\left(v_G\right)_1$ = intial velocity

Equation (7.62) states that for translational motion of a rigid body, the sum of the impulses equals the change in linear momentum of the rigid body.

7.5.2 Angular Impulse and Momentum

Now from equation (7.13)

$$\sum M_G = I_G\,\alpha$$

$$= I_G \frac{\mathrm{d}\omega}{\mathrm{d}t}$$

$$\text{or} \int M_G \,.\, \mathrm{d}t = \int_{\omega_1}^{\omega_2} I_G \,.\, \mathrm{d}\omega \qquad (7.63)$$

$$\text{or} \int M_G \,.\, \mathrm{d}t = I_G\left(\omega_2 - \omega_1\right)$$

where

ΣM_G = the sum of the moments about 'G'

I_G = the mass moment of inertia of the rigid body about 'G'

ω_2 = final angular velocity

ω_1 = initial angular velocity

Equation (7.63) states that the sum of the angular impulses on a rigid body is equal to the change in angular momentum.

Angular momentum can be described as the moment of momentum.

In the case of a rigid body **rotating about a fixed axis 'O'**, the equation relating to angular impulse and angular momentum is:

$$\Sigma \int M_O \, . \, \mathrm{d}t = I_O \left(\omega_2 - \omega_1\right) \tag{7.64}$$

where

I_O = mass moment of inertia about 'O'

$\Sigma\int M_O \, \mathrm{d}t$ = the sum of the angular impulses about 'O'

In planar motion of a rigid body, the following three equations apply:

$$m \left(v_{Gx}\right)_1 + \Sigma \int_{t_1}^{t_2} F_x \, . \, \mathrm{d}t = m\left(v_{Gx}\right)_2 \tag{7.65}$$

$$m \left(v_{Gy}\right)_1 + \Sigma \int_{t_1}^{t_2} f_y \, . \, \mathrm{d}t = m\left(v_{Gy}\right)_2 \tag{7.66}$$

$$I_G\omega_1 + \Sigma_{t_1}^{t_2} M_G \, . \, \mathrm{d}t = I_G \, . \, \omega_2 \tag{7.67}$$

where the suffices x and y refer both to the x and y directions, respectively, and the suffices 1 and 2 refer to the intial and final conditions, respectively.

Example 7.18

A solid uniform thickness disc of mass 10 kg and of external radius 0.2 m, rotates about the fixed axis 'O'. Starting from rest, the disc is subjected to a clockwise couple of 5 Nm, together with a downward force of 40 N, as shown in Figure 7.61. Determine the angular velocity of the disc after 2.5 seconds and the reaction at the discs bearing.

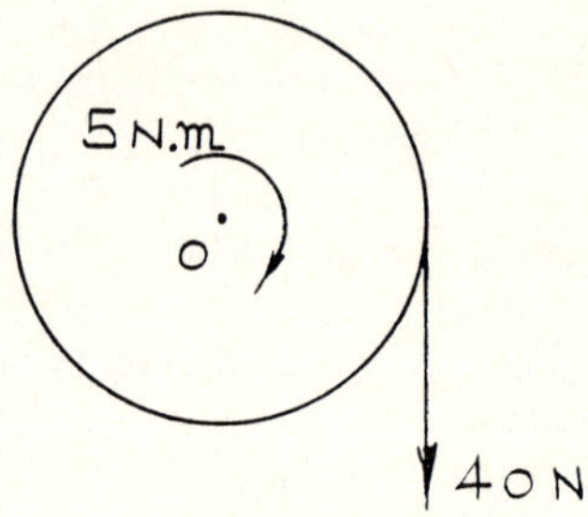

Figure 7.61

$$I_O = 10 \times 0.2^2/2 = 0.2 \text{ kg . m}^2$$

From equation (7.64)

$$I_O \times 0 + 5 \times 2.5 + 40 \times 0.2 \times 2.5 = I_O\, \omega_2$$

$$\text{or } 12.5 + 20 = 0.2\, \omega_2$$

$$\text{or } \omega_2 = 162.5 \text{ rad/s}$$

From equations (7.65) and (7.66)

$$m \times 0 + H \times 2.5 = m \times 0$$

$$\therefore H = 0$$

$$\text{and } m \times 0 + V \times 2.5 = (mg + 40) \times 2.5 + m \times 0$$

$$V = 10 \times 9.81 + 40$$

$$V = 138.1 \text{ N}$$

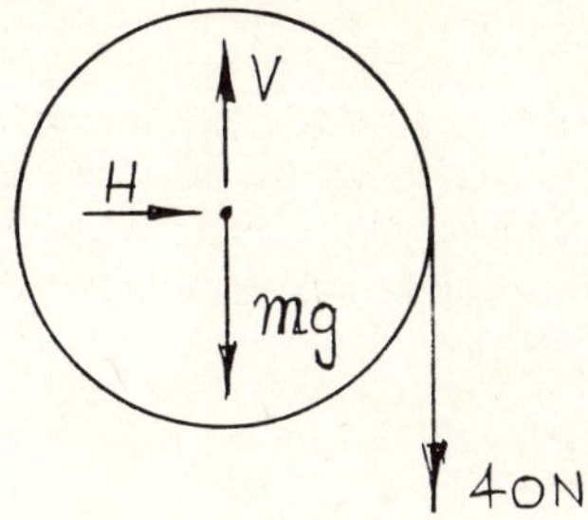

Figure 7.62

Example 7.19

A block of mass 8 kg is attached to a light cord, which is wrapped around a solid uniform thickness disc of mass 30 kg and of external radius 0.3 m. If the block is suddenly released from rest, determine the angular velocity of the disc after 2.5 seconds. Determine also the tension in the cord and the reaction at the disc's bearing.

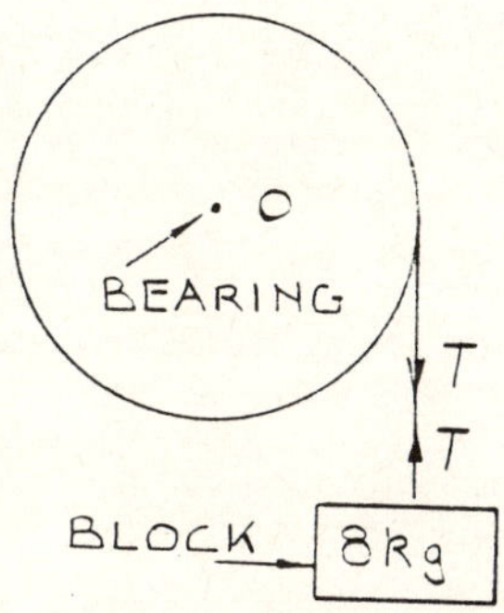

Figure 7.63

Block

From equation (7.66)

$$8 \times 0 + 8gt - Tt = 8v_2$$
$$\text{or } 196.2 - 2.5T = 8v_2 \qquad (7.68)$$

Disc

$$I_G = mr^2/2 = 30 \times 0.3^2/2 = 1.35 \text{ kg m}^2$$

From equation (7.67)

$$I_G \times 0 + T \times 0.3\,t = I_G\,\omega_2$$

$$\text{or } 0.3 \times 2.5\,T = 1.35\,\omega_2 \qquad (7.69)$$

$$\text{or } T = 1.8\,\omega_2$$

$$\text{but } v_2 = \omega_2\,r = 0.3\,\omega_2 \qquad (7.70)$$

Substituting equation (7.69) and (7.70) into equation (7.68), we get

$$196.2 - 2.5 \times 1.8\,\omega_2 = 8 \times 0.3\,\omega_2$$

$$\text{or } 196.2 = \omega_2\,(2.4 + 4.5) \qquad (7.71)$$

$$\therefore \omega_2 = 28.43 \text{ rad/s}$$

Substituting equation (7.71) into (7.69)

$$T = 51.18 \text{ N}$$

From equation (7.65)

$$H \times t = o$$

$$H = o$$

From equation (7.66)

$$mg\,t + Tt = Vt$$

$$\text{or } V = mg + T = 294.3 + 51.18$$

$$V = 345.48 \text{ N}$$

Example 7.20

A spool of mass 50 kg and with a radius of gyration of 0.3 m, has wrapped around its central hub, a light cord. If a horizontal force of magnitude '*P*' = (7 + 0.8 *t*) N, is applied to the cord, where '*t*' is in seconds, determine the angular velocity of the spool after a period of 4 seconds. Assume that no slipping takes place and that the spool is initially at rest.

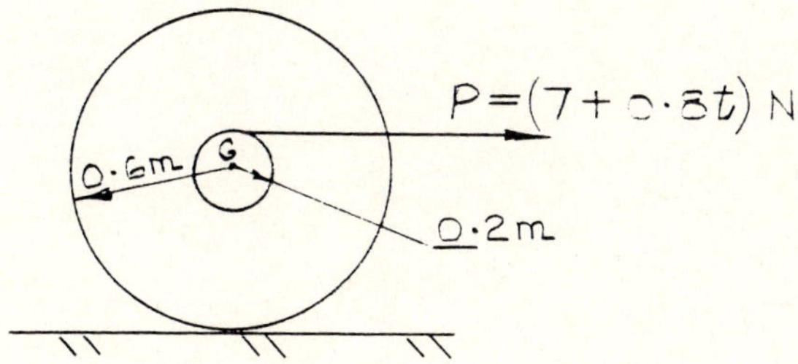

Figure 7.64

$I_G = mk^2 = 50 \text{ x } 0.3^2 = 4.5 \text{ kg . m}^2$

The freebody diagram is shown in Figure 7.65.

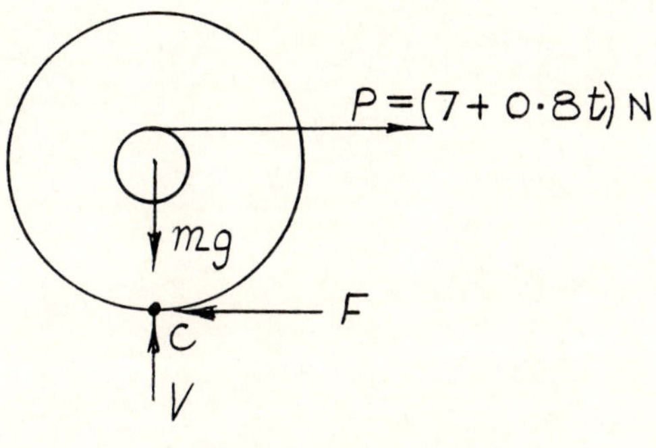

Figure 7.65 Free-body Diagram

From equation (7.65)

$$m \text{ x } 0 + \int_o^4 P \text{ . d}t - \int_o^4 F \text{ . d}t = m \left(v_{Gx}\right)_2$$

$$\int_o^4 (7 + 0.8t) \text{ . d}t - \int F \text{ . d}t = 50 \text{ . } \left(v_{Gx}\right)_2$$

$$\text{or } \left[7t + 0.8t^2/2\right]_o^4 - \int F \text{ . d}t = 50 \left(v_{Gx}\right)_2$$

$$\text{or } 34.4 - \int F \,.\, \mathrm{d}t = 50\,(v_{Gx})_2$$

$$\text{or } \int F \,.\, \mathrm{d}t = 34.4 - 50\,(v_{Gx})_2 \qquad (7.72)$$

From equation (7.67)

$$I_G \times 0 + \int_o^4 P \times 0.2\,\mathrm{d}t + \int F \times 0.6\,\mathrm{d}t = I_G \,.\, \omega_2$$

$$\text{or } 34.4 \times 0.2 + 0.6 \int F \,.\, \mathrm{d}t = 4.5\,\omega_2 \qquad (7.73)$$

$$\text{or } 6.88 + 0.6 \int F \,.\, \mathrm{d}t = 4.5\,\omega_2$$

Since the spool does not slip, the instantaneous centre of zero velocity is at 'C'.

$$\therefore\ (v_{Gx})_2 = \omega_2 \times 0.6 \qquad (7.74)$$

Substituting equation (7.72) into equation (7.73), we get

$$6.68 + 0.6\left[34.4 - 50\,(v_{Gx})_2\right] = 4.5\,\omega_2 \qquad (7.75)$$

Substituting equation (7.74) into equation (7.75), we get

$$6.88 + 20.64 - 30\,\omega_2 \times 0.6 = 4.5\,\omega_2$$

$$\text{or } 27.52 = \omega_2\,(4.5 + 18)$$

$$\text{or } \omega_2 = 1.22 \text{ rad/s (cw)}$$

7.5.3 Centre of Percussion

If a cricket bat or a baseball bat is used to strike a ball, then if the ball strikes the cricket bat or baseball bat at its centre of percussion, the batsman will not experience any reaction on his/her hands due to this action. The result of this is usually a 'six' in cricket or a 'home run' in baseball.

Example 7.2.1

Determine the position of the centre of percussion for the uniform section rod of Figure 7.66, assuming that it is gripped at the position 'A'.

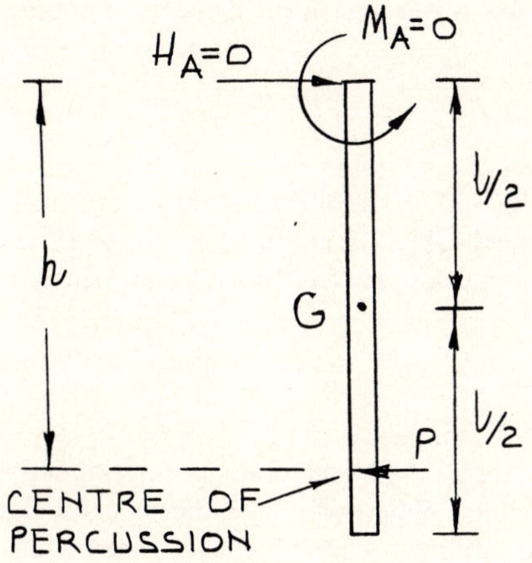

Figure 7.66

From equation (7.67)

$$\int P \, . \, dt \, . \, h = I_G \, \omega + m \, v_{Gx} \, . \, l/2 \tag{7.76}$$

From equation (7.65)

$$\int P \, . \, dt = m \, v_{Gx} \tag{7.77}$$

Substituting equation (7.77) into equation (7.76), we get

$$m \, v_{Gx} \, . \, h = I_G \, \omega + m \, v_{Gx} \, . \, l/2$$

$$\text{or } h = \frac{I_G \, \omega}{m \, v_{Gx}} + l/2$$

$$\text{but } I_G = ml^2/12$$

$$\text{and } v_{Gx} = \omega l/2$$

$$\therefore h = l/6 + l/2$$

$$h = 2l/3$$

Problems 7.6

1. A grinding wheel is being driven directly by a shaft from an electric motor at a speed of 2,000 rpm, when the electric motor is switched off. If the combined mass of the grinding wheel and the rotor is 3 kg and their combined radius of gyration is 5 cm, determine the frictional torque of the system, if the system comes to rest in 1 minute.

   [M_G = 0.0262 N.m]

2. A flywheel of mass 2,000 kg and of a radius of gyration of 1 m is rotating at a speed of 150 rpm. If the flywheel takes 10 minutes to come to rest from this speed, determine the average frictional torque exerted by the bearings.

   [M_G = 52.33 Nm]

3. A ball in a ten pin bowling game is projected along a horizontal surface, on which it initially skids, without rotating. Determine an expression for the time taken for the ball to start rolling without sliding. Determine also expressions for the linear and angular velocity at this time 't_2'. Assume that μ_k is the coefficient of kinetic friction.

   [$t_2 = 2\,v_1/(7\mu_k\,g)$; $\omega_2 = 5\,v_1/(7r)$; $v_2 = 5\,v_1/7$]

4. The striking arm of a Charpy testing machine is to be designed so that the centre of percussion is in the position shown in Figure 7.67. Determine an expression for r_p in terms of r_G and k_G.

   [$r_p = r_G + k_G^2/r_G$]

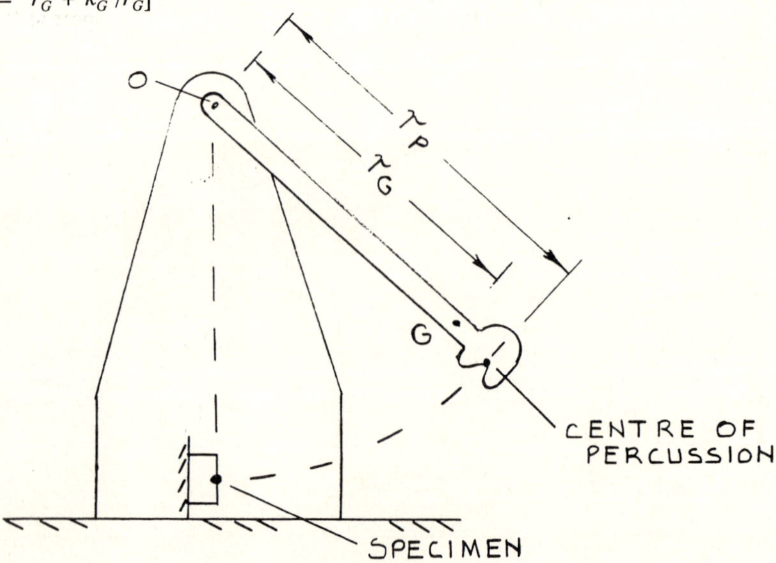

Figure 7.67

5. A wooden artefact, which has the appearance of a large mallet, hangs vertically downwards from a pin 'O', as shown in Figure 7.68. If a bullet of mass 25 grammes is fired horizontally into the artefact with a velocity of 1600 km/h, determine the resulting angular velocity of the artefact when the bullet just penetrates it. The mass of the artefact may be assumed to be 1 kg and its radius of gyration about 'O' may be assumed to be 1.2 m.

[ω = 10.45 rad/s]

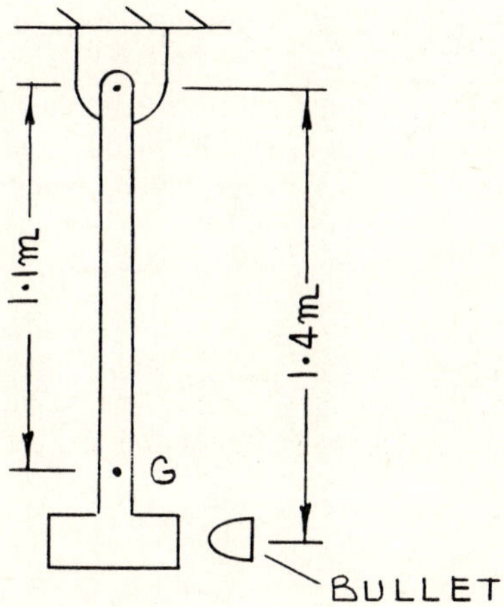

Figure 7.68

6. If in the previous problem, the bullet takes 2 x 10^{-4} seconds to imbed into the head of the artefact, determine the horizontal reaction at the pinned support at 'O'.

[H = -4.13 kN (to the left)]

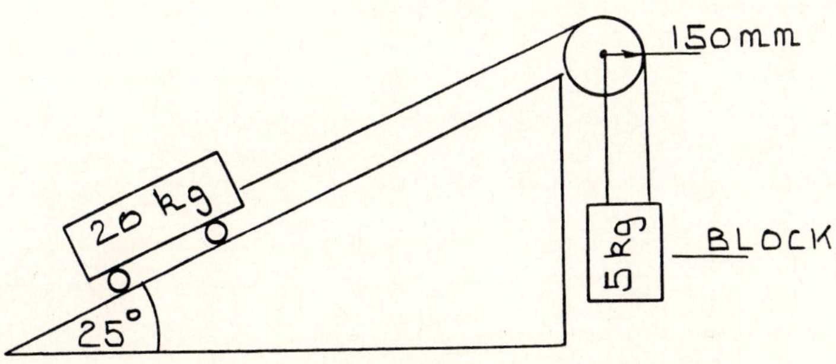

Figure 7.69

7. A mass of 20 kg, which is on wheels, is to be hauled up an incline by the method shown in Figure 7.69. If the block of 5 kg is suddenly released from rest, determine the time taken for the 20 kg mass to reach a speed of 1.8 m/s. The mass of the pulley at the top of the incline can be assumed to be 10 kg with a radius of gyration of 150 mm, but all other items can be assumed to be of negligible mass. Neglect frictional effects.

[t = 0.809 seconds]

8. A long slender rod AB of uniform section and of mass 3 kg is designed to rotate about the pin 'A', as shown in Figure 7.70. If a ball of mass 1 kg travelling horizontally with a velocity of 10 m/s, strikes the rod at mid-span, determine the angular velocity of the rod and the velocity of the ball immediately after impact. Assume that the coefficient of restitution, namely e = 0.6.

[ω_2 = 6.4 rad/s; v_{B2} = -2.8 m/s (to the left)]

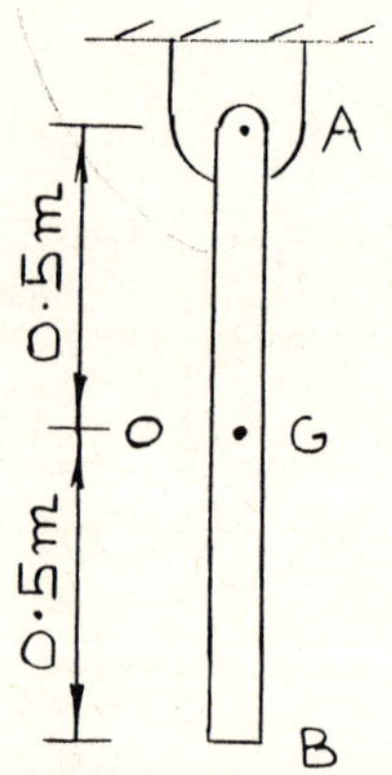

Figure 7.70

CHAPTER EIGHT

Gyroscopic Theory

8.1 INTRODUCTION

The dynamic behaviour of a gyroscope is a very interesting and important topic. A typical gyroscope is shown in Figure 8.1, where the gyroscope is free to rotate at great speed about the 'vertical' axis shown and it is free to precess and nutate at much lower velocities, about the 'horizontal' axes shown. The behaviour of a gyroscope is also shown in Figure 8.5, with the aid of a spinning top.

Figure 8.1 A 'Free' Gyroscope

Gyroscopes are important as navigational aids, because the spin axis of a 'free' gyro will always point northwards. Gyros are also used as stabilisers for ships, because a gyro can resist the rolling motion of a ship. The gyroscopic effect can be significant on the bearings of the gas turbines on aircraft and ships, when the aircraft or ship turns, while the turbines are spinning at large angular velocities.

To demonstrate gyroscopic action, consider the disc 'A' spinning about its 'z' axis with an angular velocity of ω_z, while at the same time, the disc and rod precess about the ball joint 'B' with an angular velocity of Ω_y. The effect of this combined motion will result in a moment M_x about the 'x' axis, as shown in Figure 8.2. To study gyroscopic motion, it is convenient to use fixed global X - Y - Z axes, together with moving local x - y - z axes, where the latter refer to the directions of the principal moments of inertia of the system. The symbol 'G' represents the position of the centre of gravity of the disc/rod system.

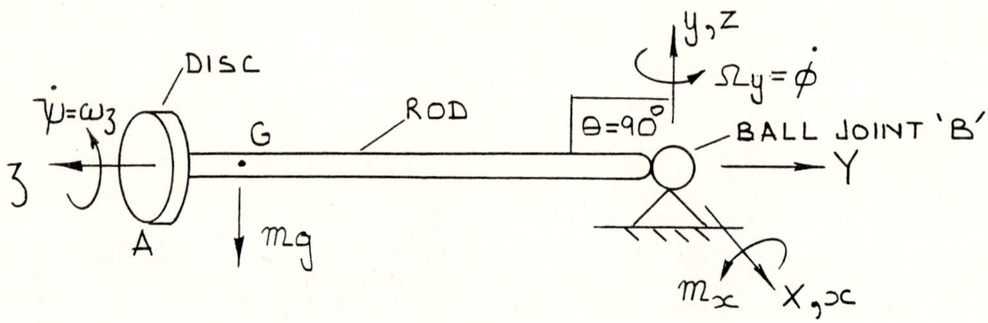

Figure 8.2

In this chapter, it will be assumed that $\omega_z >> \Omega_y$.

8.2 ANGULAR MOMENTUM OF A RIGID BODY IN THREE DIMENSIONS

The angular momentum of a rigid body in three dimensions about its centre of gravity is given by

$$\overline{H_G} = \sum \overline{r_{Gi}} \times \overline{v_{Gi}} \cdot \Delta m \tag{8.1}$$

where

$\overline{r}_{Gi}$ = the position vector of a particle 'P' from the centre of gravity 'G'

$\overline{v}_{Gi}$ = the velocity vector of the particle 'P'

$\bar{H}_G$ = angular momentum of the body about 'G'

Δm = the elemental mass of the particle 'P'

i refers to the ith particle 'P'

but $\overline{v_G} = \overline{\omega} \times \overline{r_{Gi}}$

$$\therefore \overline{H_G} = \sum \overline{r_{Gi}} \times \left(\overline{\omega} \times \overline{r_{Gi}}\right) \Delta m \tag{8.2}$$

As the body translates and rotates, it will be convenient to consider fixed global axes X - Y - Z, together with moving local axes x - y - z, where the latter refer to the directions of the principal mass moments of inertia, as shown in Figure 8.3.

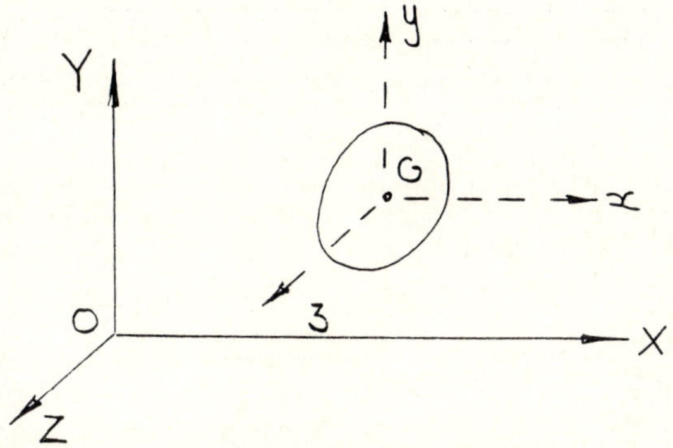

Figure 8.3

Hence, the component of angular momentum in the local x direction, can be seen to be given by the expression:

$$H_x = \sum \left[y_i \left(\overline{\omega} \times \overline{r_{Gi}}\right)_z - z_i \left(\overline{\omega} \times r_{Gi}\right)_y\right] . \Delta m$$

$$= \sum \left[y_i \left(\omega_x y_i - \omega_y x_i\right) - z_i \left(\omega_z x_i - \omega_x z_i\right)\right] \Delta m_i$$

$$= \omega_x \sum \left(y_i^2 + z_i^2\right) . \Delta m_i - \omega_y \sum x_i y_i . \Delta m_i - \omega_z \sum z_i x_i . \Delta m_i$$

$$\text{or } H_x = I_{xx} \omega_x - I_{xy} \omega_y - I_{xz} \omega_z \tag{8.3}$$

Similarly, the angular momentum in the 'y' and 'z' directions are given by:

$$H_y = -I_{yx}\,\omega_x + I_{yy}\,\omega_y - I_{yz}\,\omega_z \tag{8.4}$$

$$H_z = -I_{zx}\,\omega_x - I_{zy}\,\omega_y + I_{zz}\,\omega_z \tag{8.5}$$

where I_{xx}, I_{yy} and I_{zz} refer to the principal mass moments of inertia.

$$\text{Now } \left(\dot{\overline{H}}_G\right)_{Gxyz} = \dot{H}_x\,\bar{i} + \dot{H}_y\,\bar{j} + \dot{H}_z\,\bar{k} \tag{8.6}$$

$$\text{but } \sum \overline{M_G} = \dot{\overline{H}}_G$$

where $\bar{H}_G$ = the angular momentum of the body with respect to the X - Y - Z axes

$\left(\dot{\overline{H}}_G\right)_{Gxyz}$ = the rate of change of $\overline{H_G}$ with respect to the moving $x - y - z$ axes

$$\text{or } \sum \overline{M_G} = \dot{\overline{H}}_G = \left(\dot{\overline{H}}_G\right)_{Gxyz} + \overline{\Omega} \times \overline{H_G} \tag{8.8}$$

where $\bar{\Omega}$ = the angular velocity of the x - y - z axes.

Similarly

$$\sum M_o = \dot{\overline{H}} = \left(\dot{\overline{H}}_o\right)_{oxyz} + \overline{\Omega} \times \overline{H_o} \tag{8.8}$$

where the suffix 'o' refers to the point 'O', of the original of the X - Y - Z axes (see Figure 8.3).

8.3 GYROSCOPIC MOTION

Consider the spinning top of Figure 8.4, which shows the local moving x - y - z axes and the fixed reference X - Y - Z axes. Figure 8.5 shows the three types of gyroscopic motion, namely (i) precession, (ii) nutation and (iii) spinning.

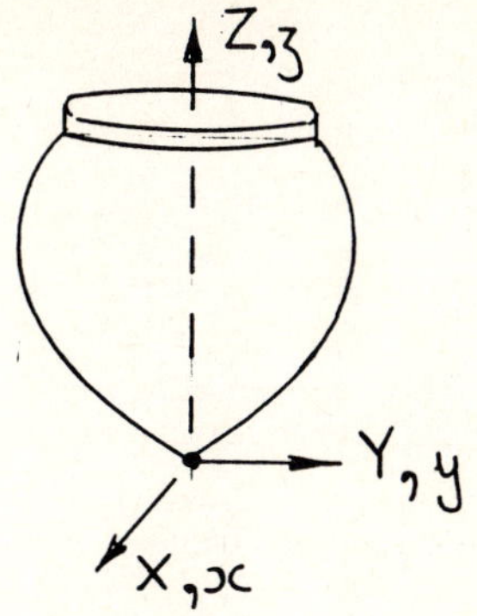

Figure 8.4

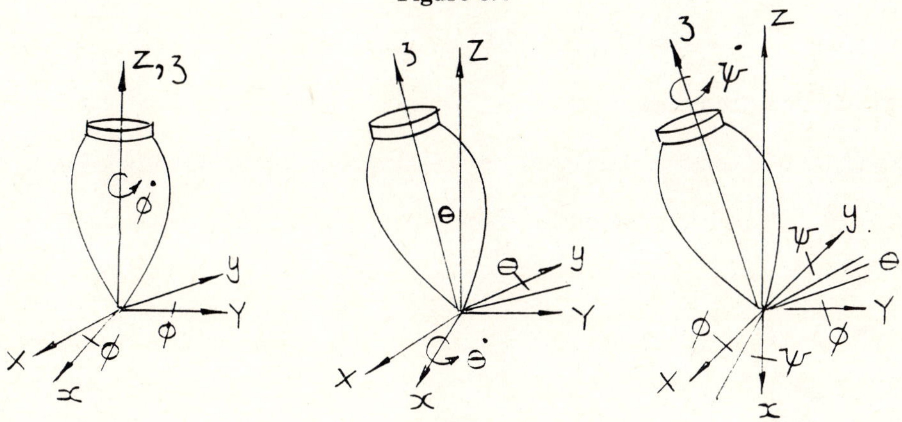

(a) Precession **(b) Nutation** **(c) Spin**

Figure 8.5

$$\overline{\omega} = \omega_x \bar{i} + \omega_y . \bar{j} + \omega_z . \bar{k}$$

$$= \dot{\theta}\, \bar{i} + (\dot{\phi} \sin\theta)\, \bar{j} + (\dot{\phi} \cos\theta + \psi)\, \bar{k}$$

As the spin component is not affected by the motion of the axes, the angular velocity vector for the x - y - z axes system is:

$$\overline{\Omega} = \Omega_x \bar{i} + \Omega_y \bar{j} + \Omega_z \bar{k}$$

$$= \dot{\theta}\, \bar{i} + (\dot{\phi} \sin\theta)\, \bar{j} + (\dot{\phi} \cos\theta)\, \bar{k}$$

Now for the top

$$I = I_{xx} = I_{yy}$$

and $I_z = I_{zz}$

$$\therefore \sum M_x = I\left(\ddot{\theta} + \dot{\phi}^2 \sin\theta \cos\theta\right) + I_z \dot{\phi} \sin\theta \left(\dot{\phi} \cos\theta + \psi\right) \tag{8.10}$$

$$\sum M_y = I\left(\ddot{\phi} \sin\theta + 2\dot{\phi}\, \dot{\theta} \cos\theta\right) - I_z \dot{\theta} \left(\dot{\phi} \cos\theta + \psi\right) \tag{8.11}$$

$$\sum M_z = I_z\left(\ddot{\psi} + \ddot{\phi} \cos\theta - \dot{\phi}\, \dot{\theta} \sin\theta\right) \tag{8.12}$$

For **steady state precession**

θ, $\dot{\phi}$ and ψ are all constant, therefore $\bar{H}_o = 0$ and equations (8.10) to (8.12) become

$$\sum M_x = -I\dot{\phi}^2 \sin\theta \cos\theta + I_z \dot{\phi} \sin\theta \left(\dot{\phi} \cos\theta + \psi\right) \tag{8.13}$$

$$\sum M_y = o$$

$$\sum M_z = o$$

Furthermore, as the spin angular velocity

In the special case, when $\theta = 90°$, as shown in Figure 8.2.

$$\sum M_x = I_z \dot{\phi}\, \psi = I_z \Omega_y \omega_z \tag{8.15}$$

where

$\dot{\phi} = \Omega_y =$ precession angular velocity

$\psi = \omega_z =$ spinning angular velocity

Example 8.1

A top of mass 0.4 kg spins with an angular velocity of $\omega_z = 80$ rad/s. If the top precesses at a constant angle of $\theta = 65°$, determine the precessional angular velocity of the top. It may be assumed that the transverse and axial mass moments of inertia are 10×10^{-4} kg m^2 and 4×10^{-4} kg m^2, respectively.

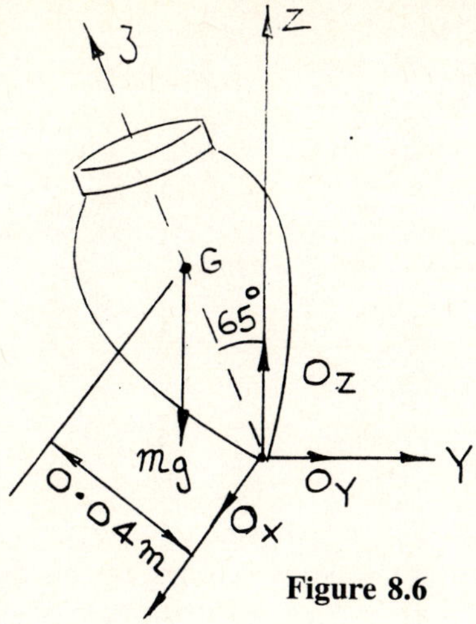

Figure 8.6

In this case, we shall use equation (8.13).

$$\text{ie} \sum M_x = -I\,\dot{\phi}^2 \sin\theta\cos\theta + I_z\,\dot{\phi}\sin\theta\left(\dot{\phi}\cos\theta + \dot{\psi}\right)$$

$$\text{where } \dot{\phi} = \Omega_y$$

$$\text{and } \dot{\psi} = \omega_z$$

$$\text{Hence} \sum M_x = mg \times 0.04 \sin 65°$$

$$= -10 \times 10^{-4}\,\dot{\phi}^2 \sin 65° \cos 65° + 4 \times 10^{-4}\,\dot{\phi}\sin 65°\left(\dot{\phi}\cos 65° + 80\right)$$

$$\text{or } 3.924 \times 0.04 \times 0.9063 = -10 \times 10^{-4}\,\dot{\phi}^2 \times 0.383$$

$$+ 4 \times 10^{-4}\,\dot{\phi} \times 0.9063\left(0.4226\,\dot{\phi} + 80\right)$$

$$0.1422 = -0.000383\,\dot{\phi}^2 + 0.000363\left(0.4226\,\dot{\phi} + 80\right)$$

$$= -0.000383\,\dot{\phi}^2 + 0.000153\,\dot{\phi}^2 + 0.029\,\dot{\phi}$$

$$\text{or } 0.00023\,\dot{\phi}^2 - 0.029\,\dot{\phi} + 0.1422 = 0$$

$$\text{or } \dot{\phi}^2 - 126.1\,\dot{\phi} + 618.3 = 0$$

$$\text{or } \dot{\phi} = 126.1 \pm \frac{\sqrt{126.1^2 - 4 \times 1 \times 618.3}}{2}$$

$$= \frac{126.1 \pm 115.9}{2}$$

$$\dot{\phi}_1 = 5.11 \text{ rad/s}$$

$$\dot{\phi}_2 = 121 \text{ rad/s}$$

The correct answer is $\dot{\phi}_1 = 5.11$ rad/s, as the system will not be able to provide sufficient kinetic energy to obtain $\dot{\phi}_2$.

Example 8.2

An artefact consists of a uniform thickness disc of mass 1.5 kg and of external radius 60 mm, spinning about the end 'A' of a rod AB, as shown in Figure 8.7.

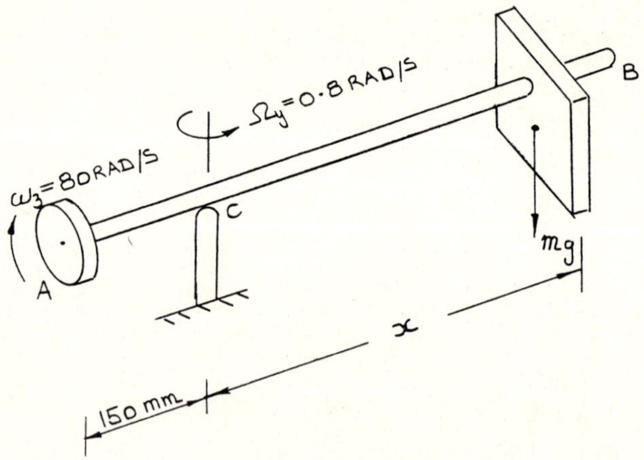

Figure 8.7

If the spin velocity of the disc is 80 rad/s, determine the distance 'x' that a mass of 3 kg has to be placed to cause the system to process with an angular velocity of $\Omega_y = 0.8$ rad/s.

$I_z = mr^2/2 = 1.5 \times 0.06^2/2 = 0.0027 \text{ kg m}^2$

Since the precession is steady, we can use equation (8.15)

ie $M_x = I_z \, \Omega_y \, \omega_z$ - See Figure 8.8.

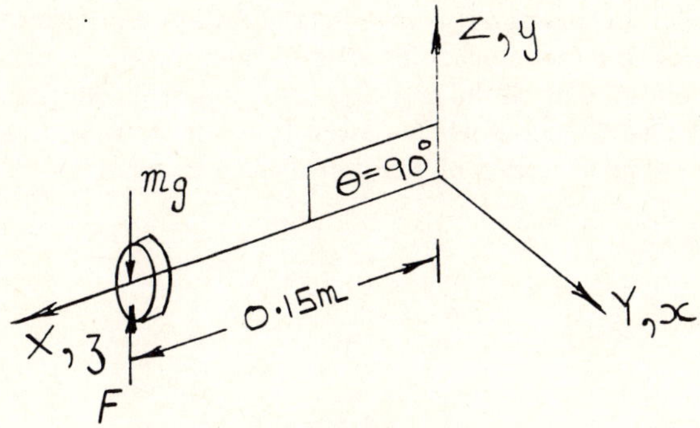

Figure 8.8

$$9.81 \times 1.5 \times 0.15 - F \times 0.15 = 0.0027 \times 0.8 \times (-80)$$

$$\text{or } 2.207 - 0.15\,F = -0.1728$$

$$F = 2.3798 \,/\, 0.15$$

$$\text{or } F = 15.865 \text{ N}$$

The free body diagram is shown in Figure 8.9.

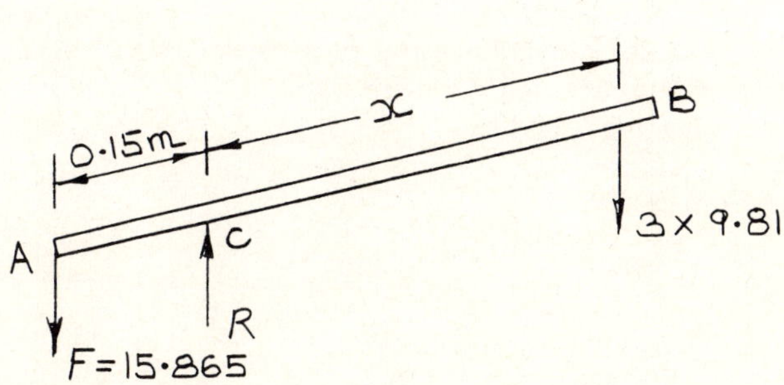

Figure 8.9

Moments about C

$$15.865 \times 0.15 = 3 \times 9.81\,x$$

$$x = 0.0809\ m$$

$$x = 80.9\ mm$$

Problems 8.1

1. A single propeller driven aircraft is travelling at a constant speed of 250 km/h, when it moves upwards in a curve of radius 90 m. The propeller's mass is 18 kg and its radius of gyration is 0.4 m. If the propeller is rotating with an angular velocity of 400 rad/s in a clockwise direction, when viewed from the front of an aircraft, determine the gyroscopic moment exerted by the propeller on its bearings.

[M_x = 888.9 Nm]

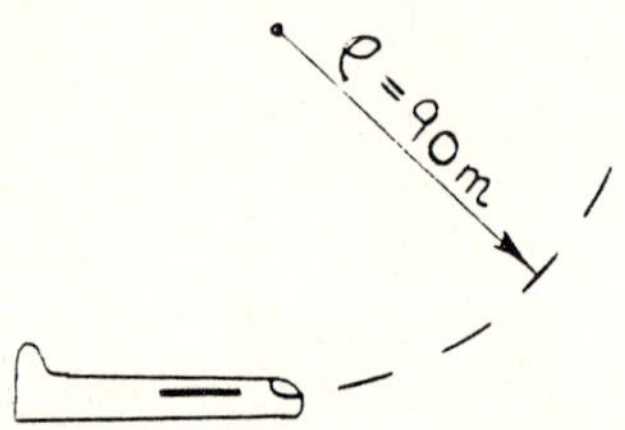

Figure 8.10

2. An air compressor consists of a turbine 'A', which drives a blower 'B' at a speed of 18000 rpm, in a clockwise direction, when viewed from left to right. The compressor lies across the aircraft, when viewed from the back of the aircraft to the front, as shown in Figure 8.11. The masses of the turbine and the blower are 3 kg and 2 kg, respectively, and their radii of gyration are 75 mm and 70 mm, respectively. If the aircraft rolls about its longitudinal axis in a clockwise direction, with an angular velocity of Ω_y = 1.5 rad/s, when viewed from the rear of the aircraft, determine the reactions of the bearing C and D.

[R_c = R_D = 603.5 N]

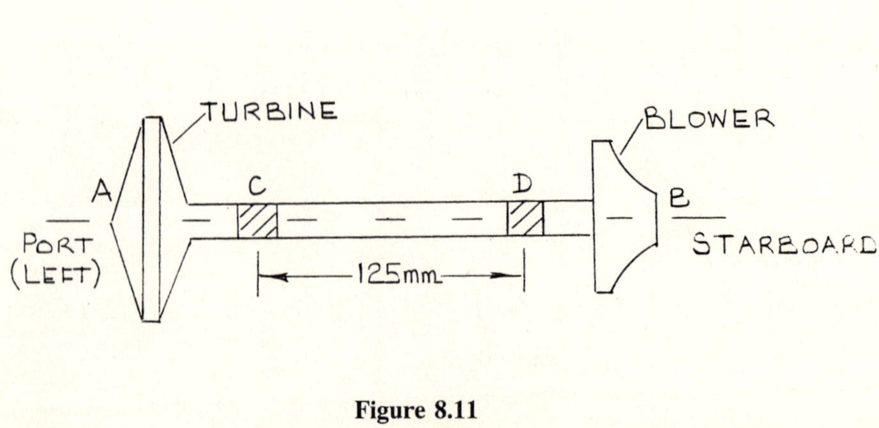

Figure 8.11

3. A uniform thickness disc of mass 40 kg and of external radius 300 mm, is attached to a rod of negligible mass. If the assembly rotates about a vertical shaft at a speed of $N = 45$ rpm, determine the normal force 'R' that acts between the wheel and the surface it is spinning on without slipping.

[$R = 525.6$ N]

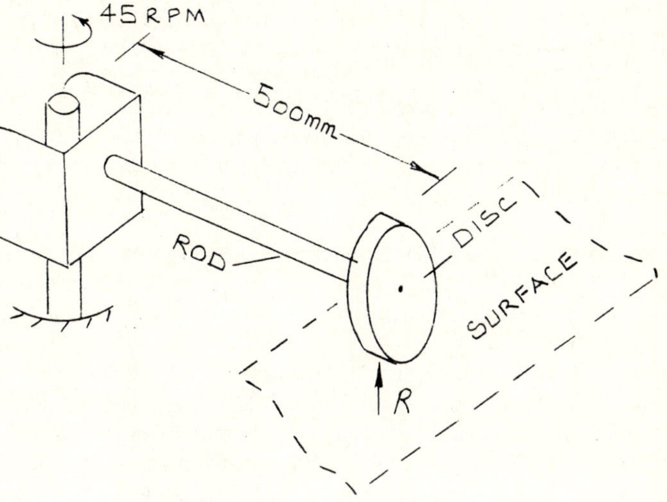

Figure 8.12

4. The rotor of a gas turbine on a ship, rotates at a speed of 6000 rpm in a counter-clockwise direction when looking from the aft end of the ship to its forward end. The rotor's mass is 900 kg and its radius of gyration is 180 mm; its centre of gravity 'G' is shown in Figure 8.13. If the ship turns to port (left, looking forward), at a speed of 20 knots, in a turning circle of radius 350 m, determine the vertical reactions at the bearings 'A' and 'B'. Will the bow (front end) rise or fall? 1 knot = 0.514 m/s

[$R_A = 5018.8$ N, $R_B = 3811.2$ N]

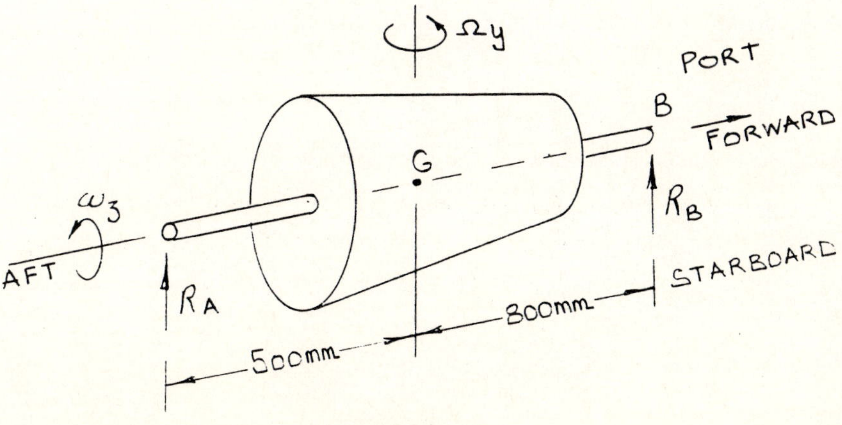

Figure 8.13

CHAPTER NINE

Free and Forced Vibrations

9.1 INTRODUCTION

The vibrations of structures and other mechanical systems is very often an undesired phenomenon. Vibrations can cause noise and other discomforts, especially to passengers and crew in transport systems. Additionally, in the case of structures, vibrations can lead to fatigue failure.

In some cases, however, vibrations are highly desirable. One such case of which is that when playing a musical instrument, while another is in the case of using vibrators to aid the satisfactory mixing of wet concrete.

Thus, a thorough knowledge of the fundamental principles of vibration is a necessary pre-requisite for most engineers and in this chapter, attempts will be made to provide such an introduction. The chapter commences with introducing and solving the one degree of freedom differential equation for the free vibration of an undamped system. The chapter then introduces viscous damping, together with forced vibrations associated with viscous damping.

Transmissibility and resonance are also defined and the chapter concludes by introducing two and multi-degrees of freedom systems. A generous supply of worked examples are provided.

9.2 UNDAMPED FREE VIBRATIONS

Prior to developing the basic theory for undamped free vibrations, it will be necessary to define *simple harmonic motion.*

9.2.1 Simple Harmonic Motion

This can be described as the motion of a particle, where the acceleration of the particle is proportional to the displacement of that particle from a mean position.

$$\text{ie } \ddot{x} + \omega^2 x = 0 \tag{9.1}$$

where x = the displacement of the particle from the mean position

ω = the circular frequency

$\ddot{x} = \dfrac{d^2x}{dt^2}$ = the acceleration of the particle

t = time

The undamped free vibrations of a one degree of freedom system can be explained with the aid of Figure 9.1, where the particle is displaced from its mean position by a distance 'x' to the right.

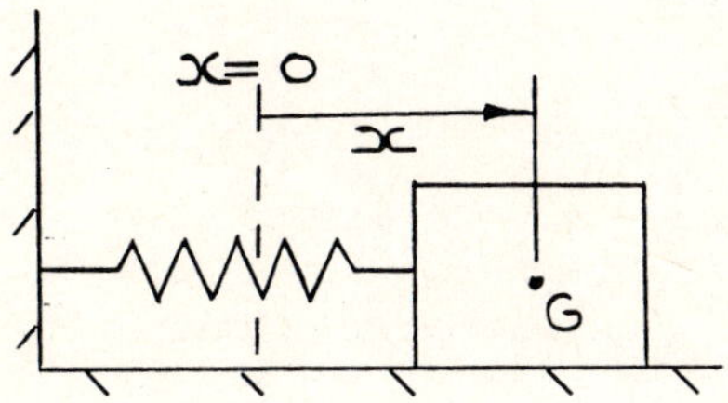

Figure 9.1

The free body diagram is shown if Figure 9.2.

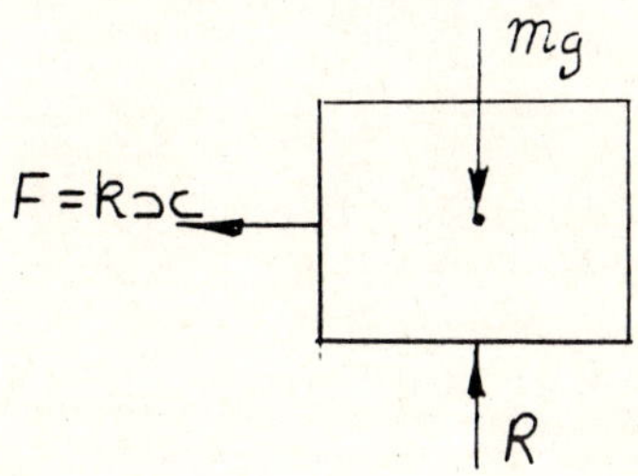

Figure 9.2 Free Body Diagram

From Newton's second law

$$F = ma = -kx$$

$$\text{or } m\,a + k\,x = 0 \qquad (9.2)$$

$$\text{or } m\,\ddot{x} + k\,x = 0$$

where

m = mass
a = $\ddot{x}$ = acceleration in the 'x' direction
k = stiffness of spring
F = force

If simple harmonic motion takes place

$$x = A \sin \omega t \tag{9.3}$$

$$\dot{x} = \omega A \cos \omega t$$

$$\text{and } \ddot{x} = -\omega^2 A \sin \omega t \tag{9.4}$$

where

A = an arbitrary constant
ω = circular frequency

Substituting equations (9.3) and (9.4) into equation (9.2), we get

$$m \cdot \left(-\omega^2 A \sin \omega t\right) + k A \sin \omega t = 0$$

$$\text{or } -\omega^2 m + k = 0$$

$$\text{or } \omega = \sqrt{k/m}$$

$$\text{ie } \omega = \omega_n = \sqrt{(k/m)} \tag{9.5}$$

= the natural frequency of the system

Consider now, the mass hanging freely from the spring of Figure 9.3(a); the free body diagram being shown in Figure 9.3(b).

(a) **(b) Free Body Diagram**

Figure 9.3

Consider the motion in the 'y' direction

where δ_{st} = static deflection of the spring = mg/k (9.6)

Applying Newton's second law

$$m\ddot{y} = mg - k(y + \delta_{st}) \tag{9.7}$$

where

y = displacement

$\ddot{y}$ = acceleration in the 'y' direction

Substituting equation (9.6) into equation (9.7)

$$-k_y = m\ddot{y}$$

$$\text{or } \ddot{y} + \frac{k}{m}y = 0 \tag{9.8}$$

$$\text{or } \ddot{y} + \omega^2 y = 0$$

To obtain a solution for equation (9.8), let

$$y = Ae^{\lambda t} \tag{9.9}$$

$$\dot{y} = \lambda Ae^{\lambda t}$$

$$\text{and } \ddot{y} = \lambda^2 Ae^{\lambda t} \tag{9.10}$$

Substituting equations (9.9) and (9.10) into equation (9.8), we get

$$\lambda^2 Ae^{\lambda t} + \omega^2 Ae^{\lambda t} = 0$$

$$\text{or } \lambda^2 + \omega^2 = 0$$

$$\text{or } \lambda = \pm j\omega \tag{9.11}$$

where

$$j = \sqrt{-1}$$

Substituting equation (9.11) into equation (9.9), we get

$$y = A_1 \cos \omega t + B_1 \sin \omega t \tag{9.12}$$

$$\text{or } y = C \sin(\omega t + \phi) \tag{9.13}$$

where

A, A_1, B_1 and C are arbitrary constants

$$C_1 = \sqrt{\left[A_1^2 + B_1^2\right]}$$

ϕ = the phase angle, as shown in Figure 9.4, where 'y' is plotted against 'ωt' = $\tan^{-1}(B_1/A_1)$

τ = time period for one cycle to take place = $2\pi/\omega$

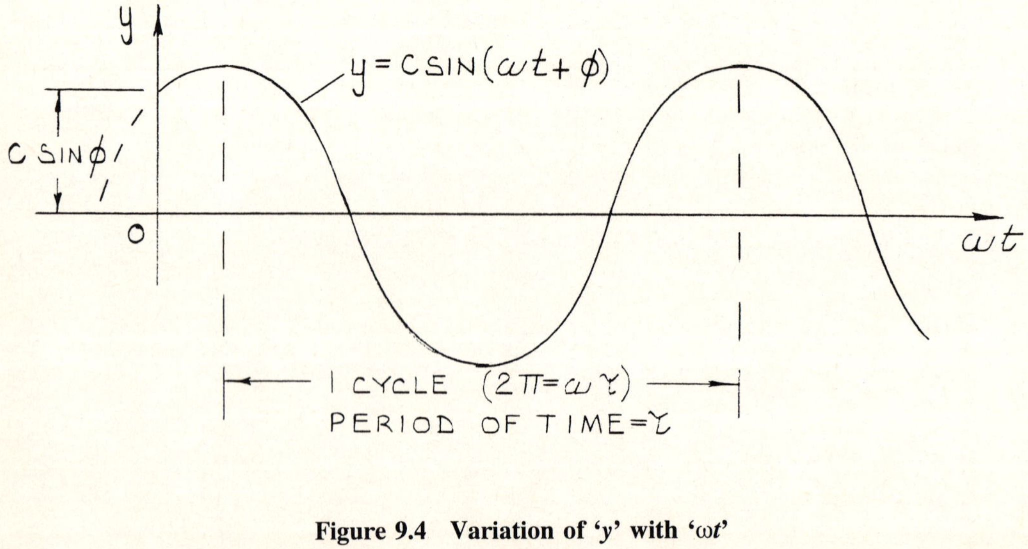

Figure 9.4 Variation of 'y' with 'ωt'

The frequency of vibration in Hz (Hertz or cycles/s) is given by:

$$f = \frac{1}{\tau} = \frac{\omega}{2\pi}\ Hz \tag{9.14}$$

$\therefore$ for an undamped one degree of freedom system $f = \dfrac{1}{2\pi}\sqrt{k/m}$ (9.15)

Example 9.1

Working from first principles determine the period of vibration for a simple pendulum, where the mass of the cord can be neglected.

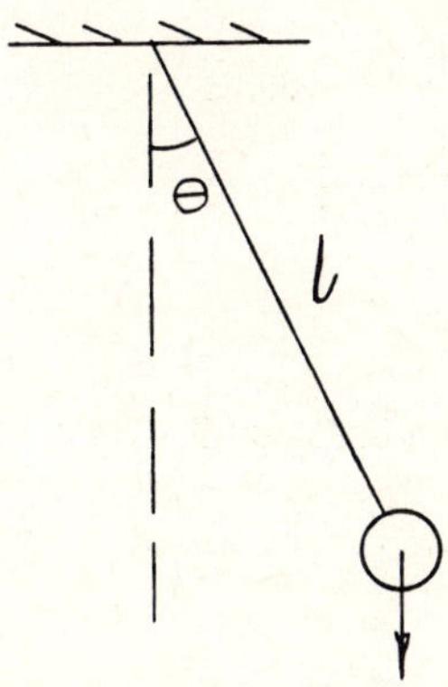

Figure 9.5 Simple Pendulum

The free body diagram is shown in Figure 9.6.

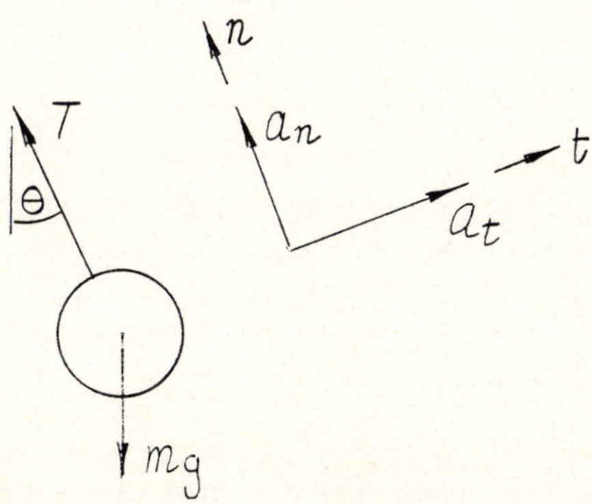

Figure 9.6 Freebody Diagram

Applying Newton's second law in the 't' direction,

$$-mg \sin\theta = ma_t = ml\ddot{\theta} \qquad (9.16)$$

As deflections are small, $\sin\theta = \theta$

Hence, equation (9.16) becomes:

$$-mg\,\theta = ml\ddot{\theta}$$
$$\text{or } \ddot{\theta} + (g/\ell)\,\theta = 0 \qquad (9.17)$$

where $\ddot{\theta}$ = angular acceleration.

Equation (9.17) is of the form

$$\ddot{x} + \omega^2 x = 0$$
$$\therefore \omega = \sqrt{g/\ell}$$

$$\text{ie } \omega_n = \sqrt{(g/\ell)} \qquad (9.18)$$

but τ = time period over one cycle

$$= 2\pi / \omega_n \qquad (9.19)$$

$$\text{or } \tau = 2\pi\sqrt{(\ell / g)}$$

From equation (9.19), it can be seen that if the periodic time of a grandfather clock is to be made longer, (ie the clock is running too fast), 'ℓ' must be made larger to slow down the clock.

Example 9.2

An 'L' shaped rod, namely ABC, which carries a mass of 7 kg at A, is pinned at B, and is supported by a horizontal spring at C. If the stiffness of the spring = 500 N/m and the masses of the rod and spring are negligible, determine the natural frequency of vibration of the system.

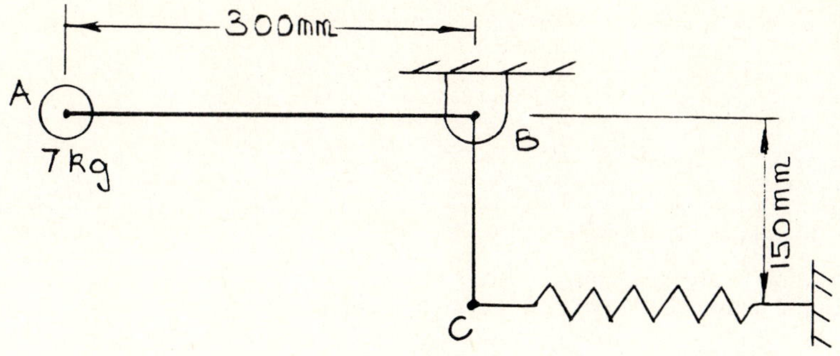

Figure 9.7

The free body diagram is shown in Figure 9.8, where the rod has been initially rotated by an angle 'θ' in a clockwise direction.

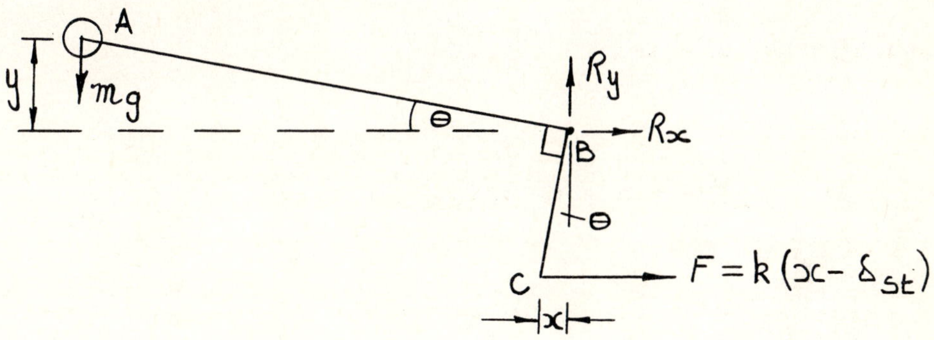

Figure 9.8 Free Body Diagram

To eliminate the unknown reactions 'R_x' and 'R_y', it will be convenient to take moments about 'B'.

$$F \times 0.15 + mg \times 0.3 = -m\, a_y \times 0.3$$

$$\text{but } F = k\left(x - \delta_{st}\right)$$

$$\therefore\ k\left(x - \delta_{st}\right) \times 0.15 + 0.3\, mg = -0.3\, m\, a_y \tag{9.20}$$

but from **initial equilibrium** considerations,

$$k \cdot \delta_{st} \times 0.15 = 0.3\, mg \tag{9.21}$$

where δ_{st} = initial static deflection of the spring at 'C'.

$$\text{Now } x = 0.15 \tan\theta = 0.15\,\theta, \text{ for small deflections} \tag{9.22}$$

$$\text{and } a_y = \ddot{y} = 0.3\,\ddot{\theta} \tag{9.23}$$

where $\ddot{\theta}$ = angular acceleration

Substituting equations (9.21) to (9.23) into equation (9.20), we get

$$k\,(0.15)^2\,\theta = -5\,(0.3)^2\,\ddot{\theta}$$

$$\text{or } 500 \times 0.0225\,\theta = -0.45\,\ddot{\theta}$$

$$\text{or } \ddot{\theta} + 25\,\theta = 0$$

which is of the form:

$$\ddot{\theta} + \omega^2\,\theta = 0$$

$$\text{ie } \omega = \sqrt{25} = 5 \text{ rad/s}$$

$$\text{or } \omega_n = 5 \text{ rad/s}$$

$$\text{and } f = \omega_n / 2\pi = 0.796 \text{ Hz}$$

Example 9.3

A 5 kg mass is attached to a light cord, which is wrapped around a uniform thickness disc of mass 8 kg and of external radius 0.25 m. Attached to the other end of the cord is a horizontal spring of stiffness 4000 N/m, as shown in Figure 9.9. Determine the natural frequency of the system, assuming that the cord does not slip over the disc.

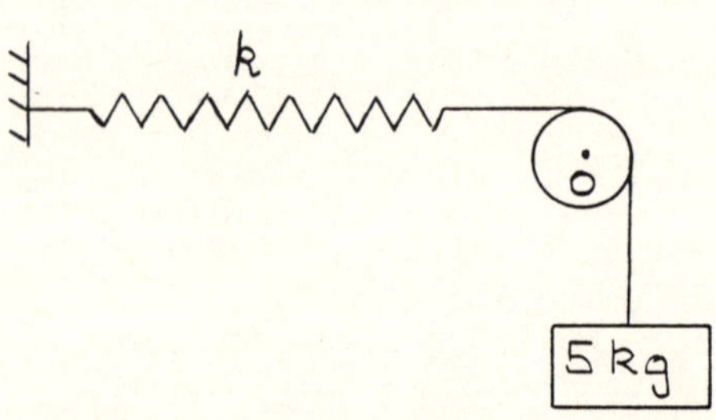

Figure 9.9

The free body diagram and the kinetic diagram are shown in Figure 9.10 (a) and (b), respectively.

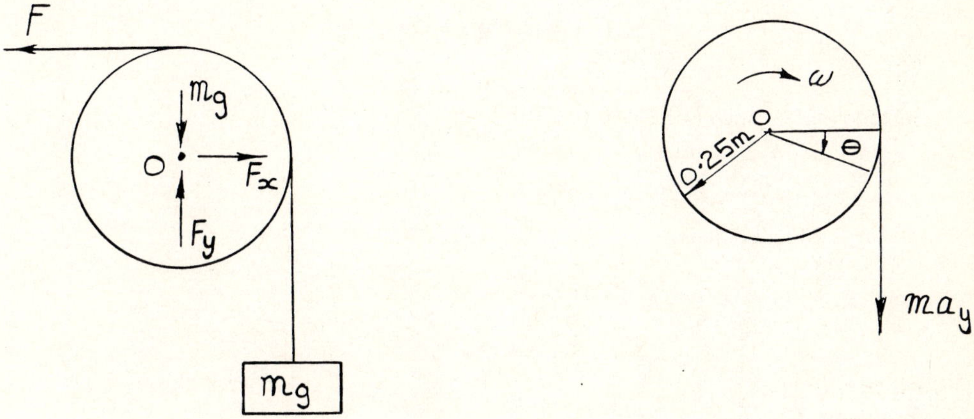

(a) Free Body Diagram (b) Kinetic Diagram

Figure 9.10

For the disc, $I_o = M r^2/2$

$$= 8 \times 0.25^2/2 = 0.25 \text{ kg.m}^2$$

$\sum M_o = \sum$ of the moments about `O'

$$= mg\,(0.25) - F\,(0.25) = I_o\,\ddot{\theta} + m\,a_y\,(0.25) \tag{9.24}$$

but $y = 0.25\,\theta$

$$\text{and } \ddot{y} = a_y = 0.25\,\ddot{\theta} \tag{9.25}$$

Additionally, from considerations of initial static equilibrium

$$F = k\,x + mg$$

$$\text{or } F = 4000 \times 0.25\,\theta + mg \tag{9.26}$$

Substituting equations (9.25) and (9.26) into equation (9.24), we get

$$mg\,(0.25) - (4000 \times 0.25\,\theta + mg)\,0.25 = 0.25\,\ddot{\theta} + 5 \times 0.25\,\ddot{\theta}\,(0.25)$$

$$\text{or } -250\,\theta = \ddot{\theta}\,(0.25 + 0.3125)$$

$$\text{or } 0.5625\,\ddot{\theta} + 250\,\theta = O$$

$$\text{or } \ddot{\theta} + 444.4\,\theta = O$$

$$\text{but } \ddot{\theta} + \omega^2\,\theta = O$$

$$\therefore \omega = \omega_n = \sqrt{444.4}$$

$$\omega_n = 21.08 \text{ rad/s}$$

$$\text{and } f = \omega_n / 2\pi = 3.355 \text{ Hz}$$

Problems 9.1

1. A mass of 5 kg is suspended from a spring of stiffness 2000 N/m, as shown in Figure 9.11. If when t = 0, $\dot{y}_o$ = 0.6 m/s, determine:

    (i) the natural frequency of the system;

    (ii) an expression for the displacement 'y', where 'y' is measured from the static equilibrium position.

    [f = 3.18 Hz; y = 0.03 sin ($\omega_n t$) m]

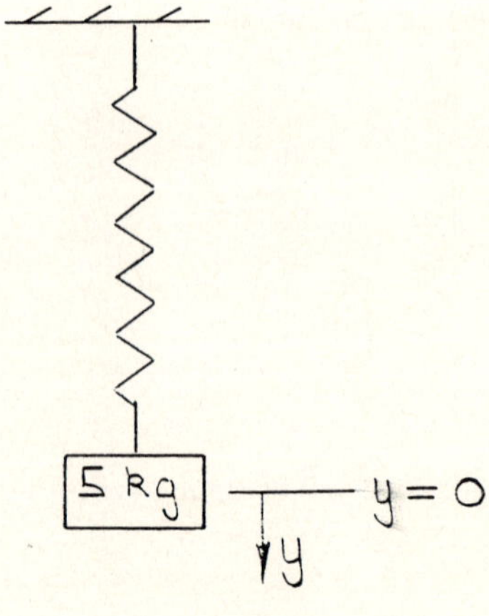

Figure 9.11

2. A rectangular block of mass 8 kg is suspended from its centre by a rod of torsional stiffness 1.2 Nm/rad, as shown in Figure 9.12.

Assuming that the mass of the rod can be neglected, determine the natural frequency of the system.

[f = 0.732 Hz]

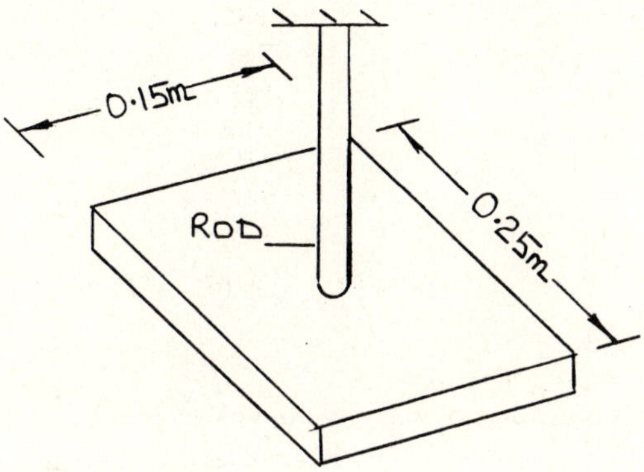

Figure 9.12

3. A mass of 80 kg is suspended symmetrically by two identical springs, each of stiffness 120 kN/m. If at $t = 0$, the downward velocity of the mass is 0.6 m/s, when the mass is passing through its equilibrium position, determine:

(i) the natural frequency of the system;

(ii) the maximum acceleration of the mass.

[f = 8.717 Hz; $\ddot{y}$ = 32.85 m/s^2]

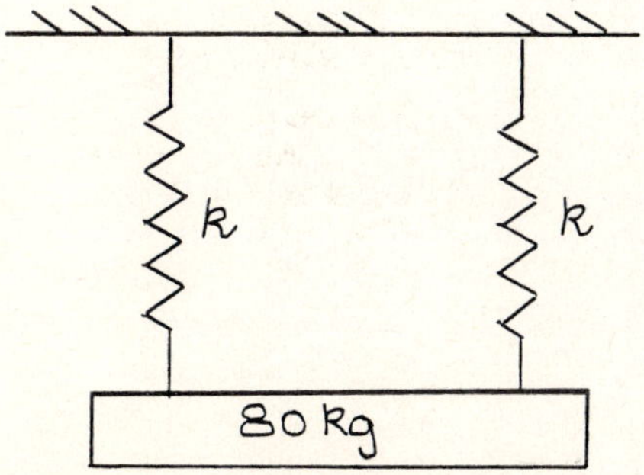

Figure 9.13

4. A circular cylindrical buoy of mass 1000 kg and of external diameter 1 m, floats vertically in salt water, which has a density of 1020 kg/m^3. Assuming that the centre of gravity of the buoy is sufficiently low to keep the buoy vertical, determine the natural frequency of the buoy.

[f = 0.446 Hz]

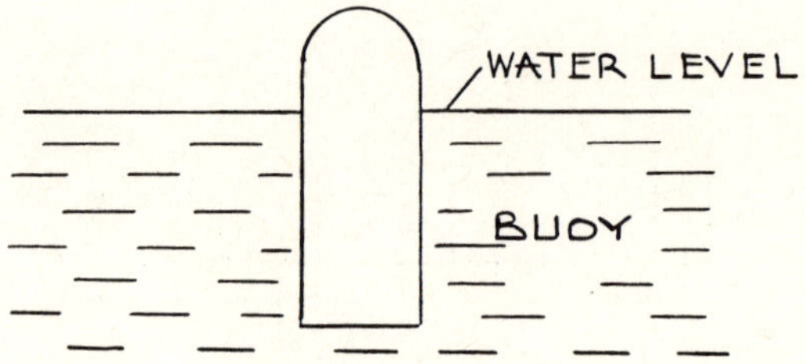

Figure 9.14

5. A uniform thickness disc of mass 'm' and of external radius 'R', is pinned on its outer circumference, as shown in Figure 9.15. Determine the natural frequency of this disc.

$$\left[f = \sqrt{[2g/3R]} \;/\; (2\pi)\right]$$

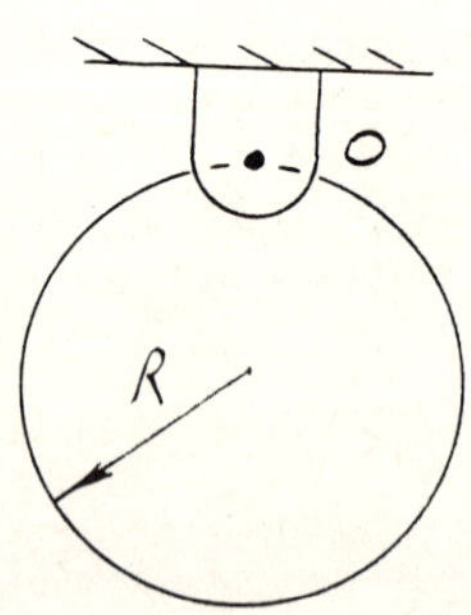

Figure 9.15

6. A uniform thickness square plate is pivoted about the point 'O' and is supported by two springs at two of its corners, as shown in Figure 9.16. Determine the natural frequency of the system in terms of 'k' and 'm'.

$$\left[f = \sqrt{[75k/(23m)]} \;/\; 2\pi\right]$$

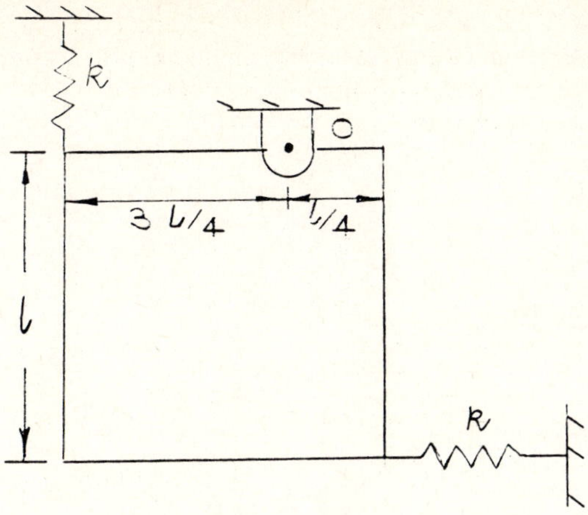

Figure 9.16

9.3 VISCOUS DAMPED FREE VIBRATION

So far, we have considered free vibration without damping, where the particle or body vibrates for evermore. In practice, however, the vibration amplitude will diminish due to damping and one such type of damping is called **viscous damping**. In viscous damping, the particle or body is assumed to vibrate in air or water or some other fluid, and the effect of the fluid is to damp out the frequency according to the expression:

$$\text{Damping force} = c\,\dot{x} \tag{9.27}$$

where $\dot{x}$ = velocity of the particle or body
c = the coefficient of damping (Ns/m)

The differential equation for one degree of freedom vibration problem, with viscous damping, can be proven with the aid of Figure 9.17, where a spring - mass - dashpot system is used.

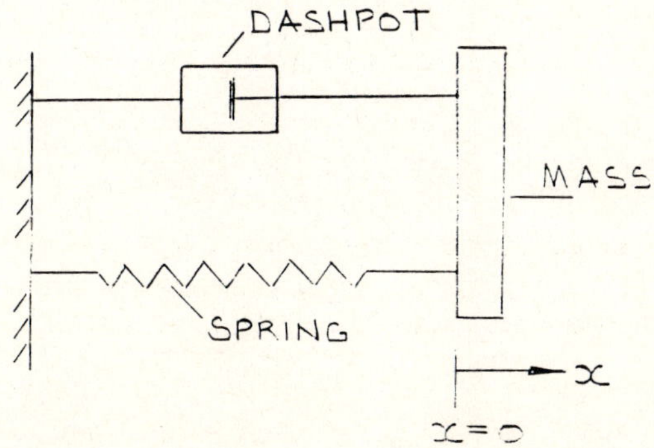

Figure 9.17 Spring - Mass - Dashpot System

In Figure 9.17, the dashpot and the spring oppose motion, so that from Newton's second law, we get,

$$-kx - c\dot{x} = m\ddot{x} \tag{9.28}$$

$$\text{or } m\ddot{x} + c\dot{x} + kx = 0 \tag{9.29}$$

To solve equation (9.29), let

$$\begin{aligned} x &= Ae^{\lambda t} \\ \dot{x} &= \lambda A e^{\lambda t} \\ \ddot{x} &= \lambda^2 A e^{\lambda t} \end{aligned} \tag{9.30}$$

Substituting equations (9.30) into equation (9.29), we get

$$\begin{aligned} m\,\lambda^2\,Ae^{\lambda t} + c\,\lambda\,Ae^{\lambda t} + k\,Ae^{\lambda t} &= 0 \\ \text{or } m\,\lambda^2 + c\,\lambda + k &= 0 \end{aligned} \tag{9.31}$$

Solving the above quadratic equation, we get,

$$\lambda_1\,,\,\lambda_2 = \frac{-c \pm \sqrt{[c^2 - 4\,mk]}}{2m}$$

$$\text{or } \lambda_1\,,\,\lambda_2 = \frac{-c}{2m} \pm \sqrt{\left[\left(\frac{c}{2m}\right)^2 - \frac{k}{m}\right]} \tag{9.32}$$

The expression for 'x' therefore becomes:

$$x = Ae^{\lambda_1 t} + Be^{\lambda_2 t} \tag{9.33}$$

Prior to describing the different types of viscous damping, it will be necessary to define the **critical damping coefficient**, namely c_c, which is obtained from the expression inside the square root sign of equation (9.32), as follows:

$$\frac{c_c^2}{4m^2} = \frac{k}{m}$$

$$\text{or } c_c = 2m\sqrt{\frac{k}{m}} \tag{9.34}$$

$$\text{but } \sqrt{\frac{k}{m}} = \omega_n$$

$$\therefore c_c = 2m\,\omega_n \tag{9.35}$$

9.3.1 For an Overdamped System

$c > c_c$

In this case, the body does not vibrate and its amplitude diminishes rapidly to its equilibrium position, without oscillating.

9.3.2 For a Critically Damped System

$c = c_c$

Here again, the body does not vibrate, but returns to its equilibrium position without oscillating, even more rapidly than an overdamped system.

9.3.3 For an Underdamped System

$c < c_c$

In this case, the body does oscillate about the equilibrium position, in values of amplitude which decrease exponentially with time, as shown in Figure 9.18; this is the usual pattern for most systems suffering vibration.

In Figure 9.18, it can be seen that

$$x = D\left[e^{-\left(\frac{c}{2m}\right)t}\sin\left(\omega_d\,t + \phi\right)\right] \tag{9.36}$$

where

D = an arbitrary constant

ω_d = the damped frequency of vibration

$$= \sqrt{\left[\left(\frac{k}{m}\right) - \left(\frac{c}{2m}\right)^2\right]} = \omega_n \sqrt{\left[1 - \left(\frac{c}{c_c}\right)^2\right]} \tag{9.37}$$

$$\text{or } \omega_d = \omega_n \sqrt{[1 - \xi^2]} \tag{9.38}$$

$\xi = \dfrac{c}{c_c}$ = the damping factor

τ_d = time period of damped vibration

$= \dfrac{2\pi}{\omega_d}$

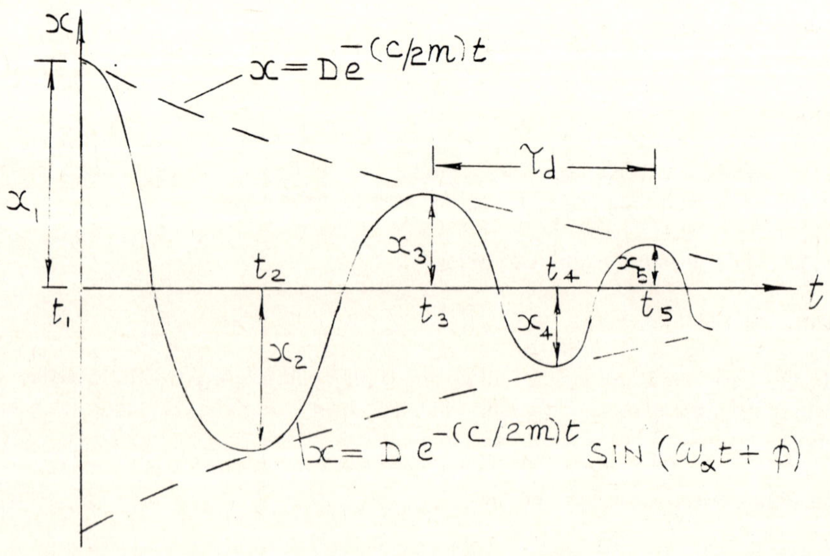

Figure 9.18

Example 9.5

Figure 9.19 shows a one degree of freedom spring - mass - dashpot system, where the mass of 5 kg is moved 0.15 m to the right and then released when $t = 0$. Assuming that $c = 15$ Ns/m and $k = 25$ N/m, determine the displacement of the mass when $t = 1.8$ s, where 't' is the time in seconds.

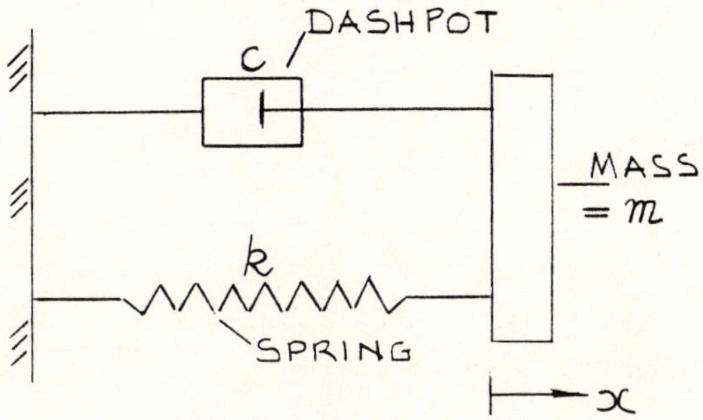

Figure 9.19 Spring - Mass - Dashpot System

From equation (9.5),

$$\omega_n = \sqrt{\frac{k}{m}} = \sqrt{\frac{25}{5}} = 2.236 \text{ rad/s}$$

From equation (9.38)

$$\omega_d = \omega_n \sqrt{\left[1 - \left(\frac{c}{c_c}\right)^2\right]}$$

but from equation (9.35)

$$c_c = 2m\,\omega_n = 2 \times 5 \times 2.236 = 22.36 \text{ Ns/m}$$

$$\therefore \omega_d = 2.236 \sqrt{\left[1 - \left(\frac{15}{22.36}\right)^2\right]}$$

$$= 2.236 \times 0.742$$

$$\omega_d = 1.658 \text{ rad/s}$$

From equation (9.36)

$$x = D\left[e^{-\left(\frac{c}{2m}\right)t} \sin\left(\omega_d\, t + \phi\right)\right]$$

$$\text{or } x = D\left[e^{-\left(\frac{15}{10}\right)t} \sin\left(1.658\, t + \phi\right)\right]$$

$$\text{and } \dot{x} = -1.5\, D\left[\, De^{-1.5t} \sin\left(1.658t + \phi\right) + 1.658\, De^{-1.5t} \cos\left(1.658t + \phi\right)\right]$$

When $t = 0,\ x = x_o = 0.15$ m

$$\therefore 0 = \mathrm{De}^0 \sin\phi = \mathrm{D} \sin\phi = 0.15$$

$$\text{or } D = 0.15 \,/\, \sin\phi \tag{9.39}$$

Similarly, @ t $= 0,\ \dot{x} = \dot{x}_o = 0$

$$\therefore \dot{x}_o = 0 = -1.5\, D \sin\phi + 1.658\, D \cos\phi \tag{9.40}$$

we get

$$0 = \frac{-1.5 \text{ x } 0.15 \sin\phi}{\sin\phi} + 1.658 \text{ x } 0.15 \cot\phi$$

$$\text{or } -0.225 = 0.2487 \cot\phi$$

$$\text{or } \frac{1}{\cot\phi} = \tan\phi = \frac{0.2487}{0.225} = 1.1053$$

$$\therefore \phi = 47.86° = 0.835\,\text{rads} \tag{9.41}$$

Substituting equation (9.41) into equation (9.39), we get

$$D = \frac{0.15}{\sin 47.86°} = 0.202\,\text{m} \tag{9.42}$$

$$\therefore x = 0.202\left[e^{(-1.5 \text{ x } 1.8)} \sin\left(1.658 + 0.835 \text{ rads}\right)\right]$$

$$= 0.202 \text{ x } 0.0672 \sin\left(142.8°\right)$$

$$= 0.01357 \text{ x } 0.6046$$

$$\text{or } x = 0.0082\,\text{m} = 8.2\,\text{mm}$$

Problems 9.2

1. A spring - mass - dashpot system is shown in Figure 9.20, where $c = 20$ Ns/m, $k = 90$ N/m and the mass $m = 1.8$ kg. Calculate the damping factor ξ.

   [$\xi = 0.786$]

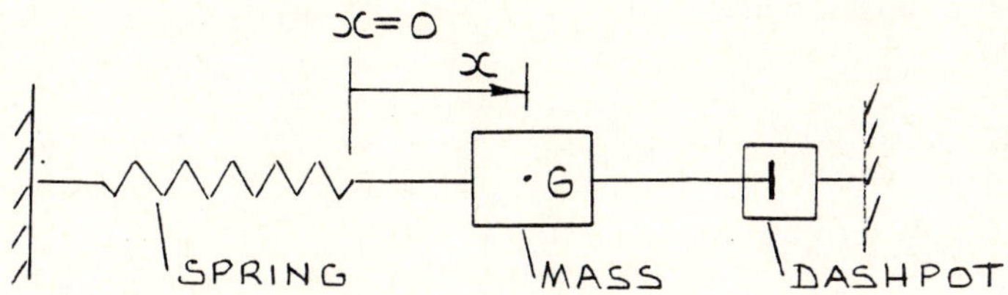

Figure 9.20

2. In Figure 9.20, the damped period of oscillation for a spring - mass - dashpot system namely $\tau_d = 0.3$ s. If $k = 800$ N/m and $m = 1.2$ kg, calculate c.

   [c 36.25 N.s/m]

3. In the spring - mass - dashpot system of Figure 9.20, it is required that $\omega_d = 0.7\ \omega_n$, what value of ξ will be required to achieve this?

   [$\xi = 0.714$]

4. If in the spring - mass - dashpot system of Figure 9.20, the mass is pulled to the right by a displacement of 'x_o', determine an expression for the displacement 'x' in terms of 't'.

   [$x = 1.617\ x_o\ \{e^{-5.556t} \sin(4.371\ t + 0.667 \text{ rads})\}$]

9.4 DAMPED FORCED VIBRATIONS

Most vibration problems are in fact damped forced vibrations and for a one degree of freedom system, the equation representing this motion is given by:

$$m\ddot{x} + c\dot{x} + kx = F_o \sin \omega t \tag{9.43}$$

where

$F_o \sin \omega t$ = the forcing function

ω = the circular frequency of the forcing function

It should be noted that the forcing function need not be sinusoidal, but for convenience, the forcing function will be assumed to be of sinusoidal form in the present text.

The solution of equation (9.43) is in two parts, a complementary function, namely x_c and a particular integral, namely x_p.

$$\text{ie } x = x_c + x_p \tag{9.44}$$

where

x_c is given by equation (9.33)

To determine x_p, let us assume that

$$x_p = x_o \sin(\omega t - \phi), \text{ so that} \tag{9.45}$$

$$\dot{x}_p = \omega x_o \cos(\omega t - \phi) \tag{9.46}$$

$$\text{and } \ddot{x}_p = -\omega^2 x_o \sin(\omega t - \phi) \tag{9.47}$$

Substituting equations (9.45) to (9.47) into equation (9.43), we get

$$\begin{aligned} &-m\,\omega^2 x_o \sin(\omega t - \phi) + c\,\omega\, x_o \cos(\omega t - \phi) \\ &+ k\, x_o \sin(\omega t - \phi) = F_o \sin \omega t \end{aligned} \tag{9.48}$$

$$\text{Let } \omega t - \phi = 0 \tag{9.49}$$

$$\text{so that } \omega t = \phi \tag{9.50}$$

Substituting equation (9.49) and (9.50) into equation (9.48), we get

$$c\,\omega\,x_o = F_o \sin\phi \tag{9.51}$$

$$\text{Now let } \omega\,t - \phi = \frac{\pi}{2}, \text{ so that} \tag{9.52}$$

$$\omega\,t = \frac{\pi}{2} + \phi \tag{9.53}$$

Substituting equations (9.52) and (9.53) into equation (9.48), we get

$$-m\,\omega^2\,x_o + k\,x_o = F_o \sin\left(\frac{\pi}{2} + \phi\right)$$

$$\text{or } -m\,\omega^2\,x_o + k\,x_o = F_o \cos\phi \tag{9.54}$$

Squaring equations (9.51) and (9.54) and adding them together, we get

$$c^2\,\omega^2\,x_o^2 + \left(k - m\,\omega^2\right)^2 x_o^2 = F_o^2\left(\sin^2\phi + \cos^2\phi\right)$$

$$= F_o^2$$

$$\text{or } F_o = \sqrt{\left[c^2\,\omega^2 + \left(k - m\,\omega^2\right)^2\right]}\; x_o \tag{9.55}$$

$$\text{or } x_o = \frac{F_o}{\sqrt{\left[c^2\,\omega^2 + \left(k - m\,\omega^2\right)^2\right]}} \tag{9.56}$$

Dividing equation (9.51) by equation (9.54), we get

$$\frac{c\,\omega\,x_o}{\left(k - m\,\omega^2\right) x_o} = \frac{F_o \sin\phi}{F_o \cos\phi}$$

$$\text{or } \tan\phi = \frac{c\,\omega}{\left(k - m\,\omega^2\right)} \tag{9.57}$$

From equation (9.5)

$$\frac{k}{m} = \omega_n^2 \tag{9.58}$$

$$\text{and } c_c = 2m\,\omega_n \tag{9.59}$$

Similarly from equation (4.9)

$$F_o = k\,\delta_o \tag{9.60}$$

where δ_o = deflection of spring due to the force F_o

$$\therefore \frac{x_o}{\left(\frac{F_o}{k}\right)} = \frac{x_o}{\delta_o} = \frac{1}{\sqrt{\left\{\left[1 - \left(\frac{\omega}{\omega_n}\right)^2\right]^2 + \left[2\left(\frac{c}{c_o}\right)\left(\frac{\omega}{\omega_n}\right)\right]^2\right\}}} \tag{9.61}$$

$$\text{and } \tan\phi = \frac{2\left(\frac{c}{c_o}\right)\left(\frac{\omega}{\omega_n}\right)}{\left[1 - \left(\frac{\omega}{\omega_n}\right)^2\right]} \tag{9.62}$$

Equation (9.61) is also known as the magnification factor, namely *MF*, because *MF* is the ratio of the vibration amplitude at t = 0 to the static deflection due to F_o.

$$\text{ie } MF = \frac{x_o}{\delta_o} = \frac{1}{\sqrt{\left[(1 - r^2)^2 + (2\,\xi\,r)^2\right]}} \tag{9.63}$$

where

$$r = \frac{\omega}{\omega_n} = \text{frequency ratio}$$

$$\xi = \frac{c}{c_c} = \text{damping factor}$$

A plot of *MF* against '*r*' is given in Figure 9.21 for different value of 'ξ'. From Figure 9.21, it can be seen that when ξ is small, the magnification factor can become quite large, especially near values of $r = 1$.

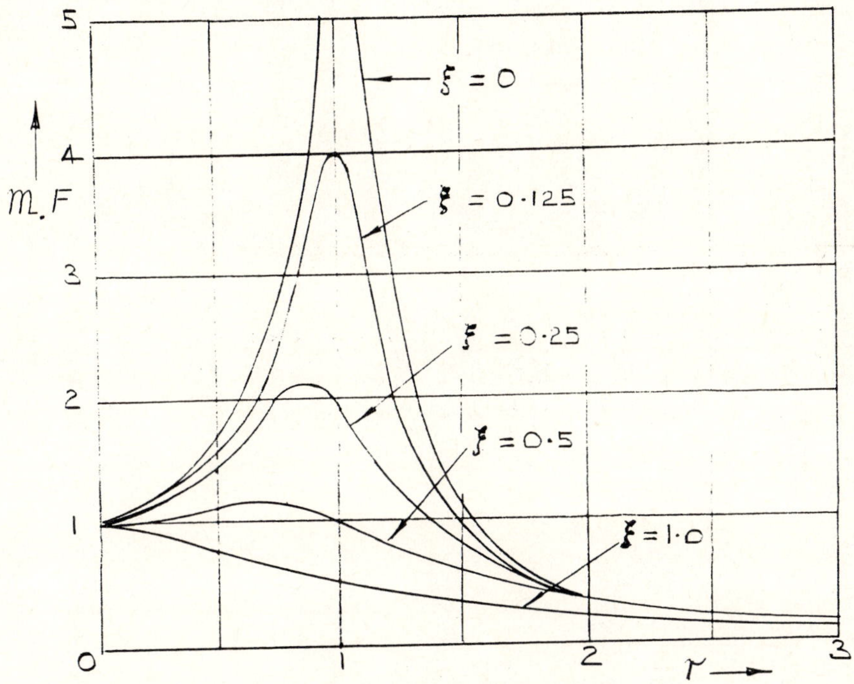

Figure 9.21 Plot of *MF* Against '*r*'

9.4.1 Other Forms of Damping

Other forms of damping include:

Rayleigh damping and structural damping. These forms of damping are beyond the scope of the present text and are described in reference [1].

9.4.2 Resonance

When the magnitude of the forcing frequency, namely 'ω' approaches the magnitude of the natural frequency, namely 'ω_n' the values of vibration amplitude, namely x_o, can become very large and the system can resonate dangerously. The result of this is that the system suffers from large displacements and this can lead to catastrophic physical collapse of the system, or in the case of a structure can lead to fatigue or other structural failure.

Example 9.6

An electric motor of mass 20 kg is symmetrically supported by four identical vertical springs, each of stiffness 150 N/m. If the rotor in the electric motor is unbalanced, so that while it is rotating with an angular velocity of 12 rad/s, the unbalanced force is equivalent to a mass of 3 kg rotating with an eccentricity of 50 mm from the axis of the rotor's shaft. If the damping factor, namely 'ξ' is 0.12, determine the amplitude of vibration of the electric motor.

$$F_o = m\,a_n = m\,\omega^2 r = 3 \times 12^2 \times 0.05 = 21.6 \text{ N}$$

$$k = 4 \times 150 = 600 \text{ N/m}$$

$$\omega_n = \sqrt{\frac{k}{m}} = \sqrt{\left(\frac{600}{20}\right)} = 5.477 \text{ rad/s}$$

$$r = \frac{\omega}{\omega_n} = \frac{12}{5.477} = 2.191$$

$$\xi = \frac{c}{c_c} = 0.12$$

From equation (9.63)

$$MF = \frac{x_o}{\delta_o} = \frac{x_o}{\left(\frac{F_o}{k}\right)}$$

$$= \frac{1}{\sqrt{\left[\left(1 - r^2\right)^2 + \left(2\,\xi\,r\right)^2\right]}}$$

$$\text{or } x_o = \frac{\frac{F_o}{k}}{\sqrt{\left[\left(1 - 2.191^2\right)^2 + \left(2 \times 0.12 \times 2.191\right)^2\right]}}$$

$$= \frac{\frac{21.6}{600}}{\sqrt{\left(14.441 + 0.2765\right)}} = \frac{0.036}{3.8367}$$

$$\text{or } x_o = 0.00938\,\text{m} = 9.38\text{ mm}$$

Problems 9.3

1. The differential equation for a one degree of freedom, forced but undamped system of vibration is given by:

$$m \ddot{x} + k x = F_o \sin \omega t$$

Determine an expression for the displacement 'x' in terms of 't'.

$$\left[x = A \sin (\omega_n t) + B \cos (\omega_n t) + \left(\frac{F_o}{k}\right) \frac{\sin (\omega t)}{1 - r^2} \right]$$

2. An electrical instrument of mass 30 kg is supported symmetrically on four identical vertical springs, each of stiffness 5000 N/m. If the foundation on which the instrument stands is subjected to a disturbance of $x_F = 0.0015 \cos 50$ and damping is negligible, determine the amplitude of the steady state motion, where x_F = the displacement at the base of the instrument.

[x_o = -0.545 mm]

3. An electrical instrument of mass 8 kg is supported symmetrically by four identical vertical springs, each of stiffness of 1200 N/m. If the base of the instrument is subjected to a maximum vertical displacement of 0.05 m at a frequency of 8 Hz, determine the amplitude of the steady state motion. Neglect damping.

[x_o = -15.6 mm]

4. An electrical motor of mass 20 kg is supported symmetrically in four identical vertical springs, each of stiffness 100,000 N/m. If the imbalance of its rotor is equivalent to a mass of 0.1 kg rotating with an eccentricity of 0.15 m from the rotor's axis of rotation, determine the steady state amplitude of vibration of the motor when the rotor is rotating at 1700 rpm when:

(i) $\xi = o$
(ii) $\xi = 0.12$

[x_o = 2.03 mm; x_o = 1.8 mm]

5. An electrical motor of mass 30 kg is supported symmetrically on four identical vertical springs, each of stiffness 120 000 N/m. If the rotor in the electrical motor is unbalanced and equivalent to a mass of 0.2 kg rotating eccentrically by a distance of 0.1 m from the centre of the rotor's shaft, determine the steady state amplitude of vibration. The rotor is rotating at a speed of 2000 rpm. Assume $\xi = 0.14$.

[x_o = 0.933 mm]

9.5 TRANSMISSIBILITY

One method of reducing the magnitude of the force transmitted to artefacts and other objects, is to support the artefact or object on springs and dampers. When this is done, the transmitted force is the sum of the spring force and the damper force, so that for viscous damping.

The transmitted force $F_t = k\,x + c\,\dot{x}$, and the transmissibility ratio, namely TR is defined as:

$$TR = \frac{F_t}{F_o}$$

$$= \frac{\sqrt{\left[1 + (2\,\xi\,r)^2\right]}}{\sqrt{\left[(1 - r^2)^2 + (2\,\xi\,r)^2\right]}} \qquad (9.64)$$

9.6 TWO AND MULTI-DEGREES OF FREEDOM SYSTEMS

Prior to considering multi-degree of freedom problems, it will be necessary to consider the undamped two degree of freedom system shown in Figure 9.22.

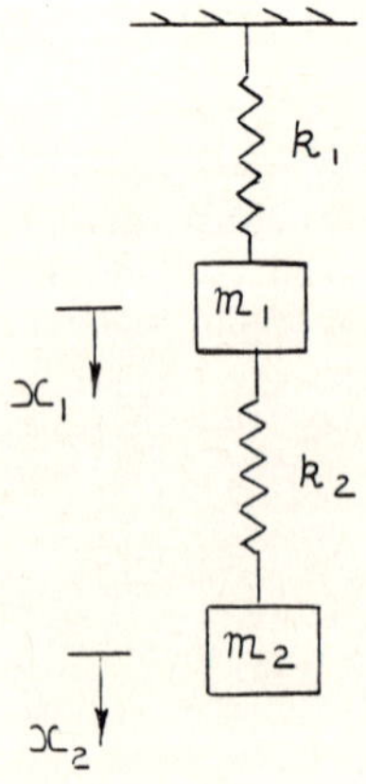

Figure 9.22

Applying Newton's 2nd law to the system shown in Figure 9.22, we get the following two simultaneous differential equations:

$$m_1 \ddot{x}_1 = -k_1 x_i - k_2 (x_1 - x_2) \tag{9.65}$$

and $m_2 \ddot{x}_2 = -k_2 (x_2 - x_1)$ (9.66)

which when rearranged becomes:

$$m_1 \ddot{x} + (k_1 + k_2) x_1 - k_2 x_2 = o \tag{9.67}$$

$$m_2 \ddot{x}_2 - k_2 x_1 + k_2 x_2 = o \tag{9.68}$$

Assuming that the motion is periodic

ie $x_1 = A \sin(\omega t + \phi)$ (9.69)

$$x_2 = B \sin(\omega t + \phi) \tag{9.70}$$

and substituting equation (9.69) and (9.70) and their appropriate derivatives into equations (9.67) and (9.68), we get

$$-m_1 A \omega^2 \sin(\omega t + \phi) + (k_1 + k_2) A \sin(\omega t + \phi) - k_2 B \sin(\omega t + \phi) \tag{9.71}$$

and $-m_2 B \omega^2 \sin(\omega t + \phi) + k_2 B \sin(\omega t + \phi) - k_2 A \sin(\omega t + \phi) = 0$ (9.72)

In matrix form equations (9.71) and (9.72) becomes:

$$\begin{bmatrix} (k_1 + k_2 - m_1 \omega^2) & -k_2 \\ -k_2 & (k_2 - m_2 \omega^2) \end{bmatrix} \begin{Bmatrix} A \\ B \end{Bmatrix} = \begin{Bmatrix} 0 \\ 0 \end{Bmatrix} \tag{9.73}$$

or

$$\begin{bmatrix} \dfrac{(k_1 + k_2)}{m_1} & \dfrac{-k_2}{m_1} \\ \dfrac{-k_2}{m_2} & \dfrac{k_2}{m_2} \end{bmatrix} - \omega^2 \begin{bmatrix} 1 & 0 \\ 0 & 1 \end{bmatrix} \begin{Bmatrix} x_1 \\ x_2 \end{Bmatrix} = \begin{Bmatrix} 0 \\ 0 \end{Bmatrix} \tag{9.74}$$

which is a standard eigenvalue problem of the form:

$$\left([k] - \lambda\,[I]\right)\{x_i\} = \{0\} \tag{9.75}$$

As the condition $\{x_i\} = \{0\}$ is not of interest, the following determinant is zero:

$$\left|\,[k] - \lambda\,[I]\,\right| = 0 \tag{9.76}$$

In this case we get

$$\begin{bmatrix} \dfrac{(k_1 + k_2)}{m_1} - \omega^2 & \dfrac{-k_2}{m_1} \\ & \\ \dfrac{-k_2}{m_2} & \dfrac{k_2}{m_2} - \omega^2 \end{bmatrix} = 0 \tag{9.74}$$

$$\text{or } \omega^4 - \omega^2 \left(\frac{k_1 + k_2}{m_1} + \frac{k_2}{m_2} \right) + \frac{k_1\,k_2}{m_1\,m_2} = o$$

Solution of the above quadratic equation, yields the following roots:

$$\omega^2 = \frac{(k_1 + k_2)}{2m_1} + \frac{k_2}{2m_2} \pm \sqrt{\left\{ \frac{1}{4} \left[\frac{(k_1 + k_2)}{m_1} + \frac{k_2}{m_2} \right]^2 - \frac{k_1\,k_2}{m_1\,m_2} \right\}} \tag{9.77}$$

9.6.1 Multi-degrees of Freedom Problems

For multi-degrees of freedom problems, equation (9.76) can be solved with the aid of a suitable computer program and a computer. For complex structures, the finite element method can be used, as described in references [1 to 3]. In references [2 and 3] **listings of computer programs** are published which can be used for solving vibration problems for complex structures. They also contain publications of eigensolvers.

9.7 REFERENCES

1. Ross C T F, 'Finite Element Techniques in Structural Mechanics', Albion Publishing, 1996.

2. Ross C T F, 'Finite Element Programmes in Structural Engineering and Continuum Mechanics', Albion Publishing, 1996.

3. Ross C T F, 'Finite Element Methods in Engineering Science', Ellis Horwood, 1990.

APPENDIX I

Simple Vector Analysis

Prior to introducing vector analysis, the rules for expanding 2 x 2 (2nd order) and 3 x 3 (3rd order) determinants will be given.

The rule for expanding second order determinants is as follows:-

$$|A| = \begin{vmatrix} a_{11} & a_{12} \\ a_{21} & a_{22} \end{vmatrix}$$

so that

D = the value of the determinant

$= a_{11}\,a_{22} - a_{12}\,a_{21}$

The method used for expanding second order determinants can be used for expanding third order determinants as follows:-

$$|A| = \begin{vmatrix} a_{11} & a_{12} & a_{13} \\ a_{21} & a_{22} & a_{23} \\ a_{31} & a_{32} & a_{33} \end{vmatrix}$$

so that

$$D = a_{11} \begin{vmatrix} a_{22} & a_{23} \\ a_{32} & a_{33} \end{vmatrix} - a_{12} \begin{vmatrix} a_{21} & a_{23} \\ a_{31} & a_{33} \end{vmatrix} + a_{13} \begin{vmatrix} a_{21} & a_{22} \\ a_{31} & a_{32} \end{vmatrix}$$

$$D = a_{11}\,(a_{22}\,a_{33} - a_{23}\,a_{32}) - a_{12}\,(a_{21}\,a_{33} - a_{23}\,a_{31}) + a_{13}\,(a_{21}\,a_{32} - a_{22}\,a_{31})$$

Expansion of determinants of order greater than 3 by the above method is extremely difficult, and other methods have to be used.

The unit vectors $\bar{i}$, $\bar{j}$ and $\bar{k}$ are shown in Figure AI.1, where

$$\bar{i} = [1 \quad 0 \quad 0]$$

$$\bar{j} = [0 \quad 1 \quad 0]$$

$$\bar{k} = [0 \quad 0 \quad 1]$$

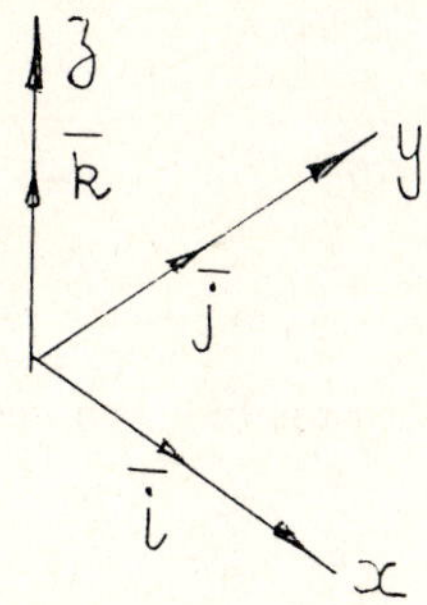

Figure AI.1 Unit Vectors

The **dot product** of the vectors $\bar{A}$ and $\bar{B}$ is as follows:

$$\bar{A} \cdot \bar{B} = [a\bar{i} \quad b\bar{j} \quad c\bar{k}] \cdot [d\bar{i} \quad e\bar{j} \quad f\bar{k}]$$

$$= ad + be + cf$$

where $\bar{i} \cdot \bar{i} = \bar{j} \cdot \bar{j} = \bar{k} \cdot \bar{k} = 1$

$$\bar{A} = [a\bar{i} \quad b\bar{j} \quad c\bar{k}]$$

$$\bar{B} = [d\bar{i} \quad e\bar{j} \quad f\bar{k}]$$

Thus $\bar{i} \cdot \bar{j} = \bar{i} \cdot \bar{k} = \bar{j} \cdot \bar{k} = \bar{j} \cdot \bar{i} = \bar{k} \cdot \bar{i} = \bar{k} \cdot \bar{j} = 0$

and $\bar{i} \cdot \bar{i} = \bar{j} \cdot \bar{j} = \bar{k} \cdot \bar{k} = 1$

The **cross product** of two vectors $\bar{A}$ and $\bar{B}$ is as follows:

$$\overline{C} = \overline{A} \times \overline{B} = \begin{vmatrix} \overline{i} & \overline{j} & \overline{k} \\ a & b & c \\ d & e & f \end{vmatrix}$$

$$\text{or } \overline{C} = \overline{i}\,(bf - ce) - \overline{j}\,(af - cd) + \overline{k}\,(ae - bd)$$

Thus

$$\overline{i} \times \overline{j} = \overline{k} \qquad \overline{j} \times k = \overline{i} \qquad \overline{k} \times \overline{i} = \overline{j}$$

$$\overline{j} \times i = -k \qquad \overline{k} \times \overline{j} = -i \qquad \overline{i} \times \overline{k} = -j$$

$$\overline{i} \times \overline{i} = \overline{j} \times \overline{j} = \overline{k} \times \overline{k} = 0$$

APPENDIX II

Mass Moment of Inertia and other Geometrical Properties

(a) Areas and Second Moments of Area

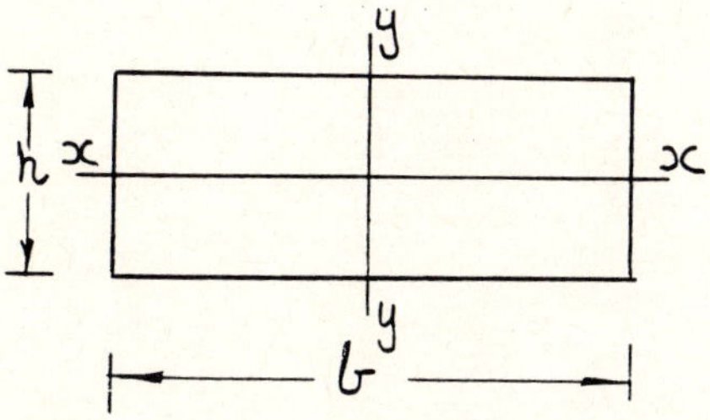

$$I_x = \frac{1}{12} bh^3$$

$$I_y = \frac{1}{12} hb^3$$

$$A = bh$$

Rectangular Area

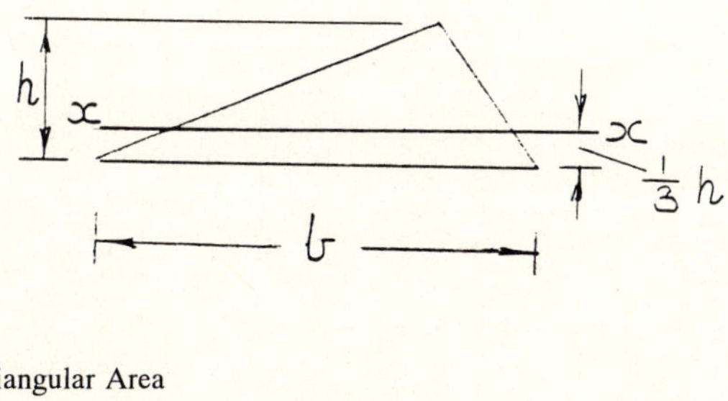

$$I_x = \frac{1}{36} bh^3$$

$$A = \frac{1}{2} bh$$

Triangular Area

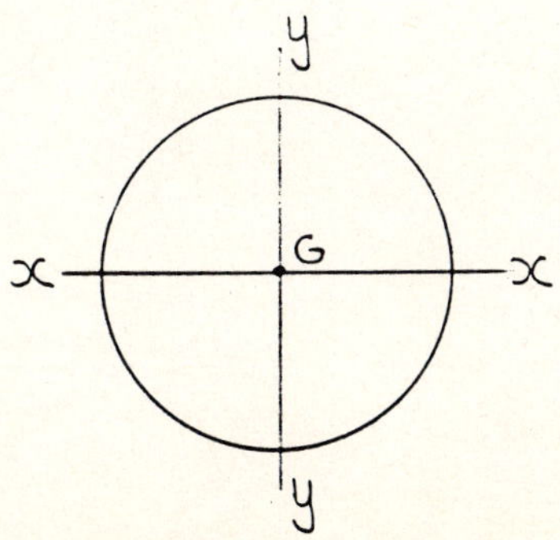

$$I_x = \frac{1}{4} \pi r^4$$

$$I_y = \frac{1}{4} \pi r^4$$

$$A = \pi r^2$$

Circular Area

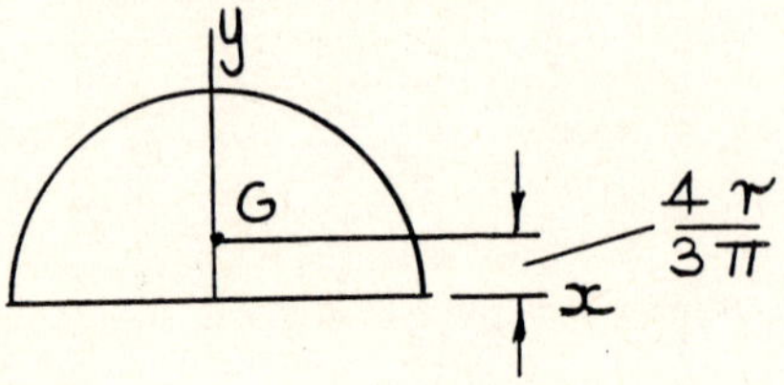

$$I_x = \frac{1}{8}\pi r^4$$

$$I_y = \frac{1}{8}\pi r^4$$

$$A = \frac{\pi r^2}{2}$$

Semicircular Area

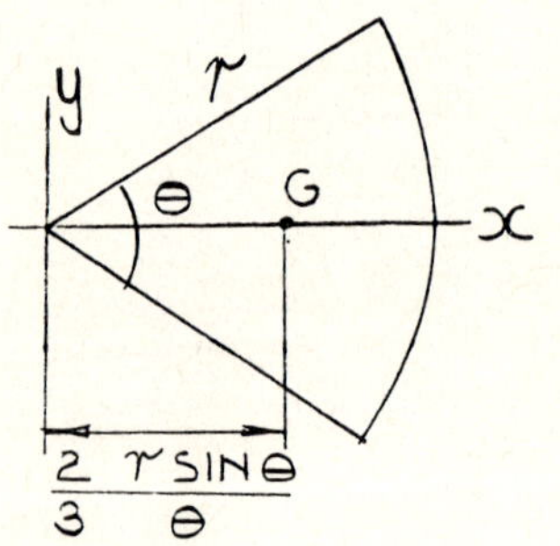

$$I_x = \frac{1}{4}r^4\left(\theta - \frac{1}{2}\sin 2\theta\right)$$

$$I_y = \frac{1}{4}r^4\left(\theta + \frac{1}{2}\sin 2\theta\right)$$

$$A = \theta r^2$$

Circular Sector Area

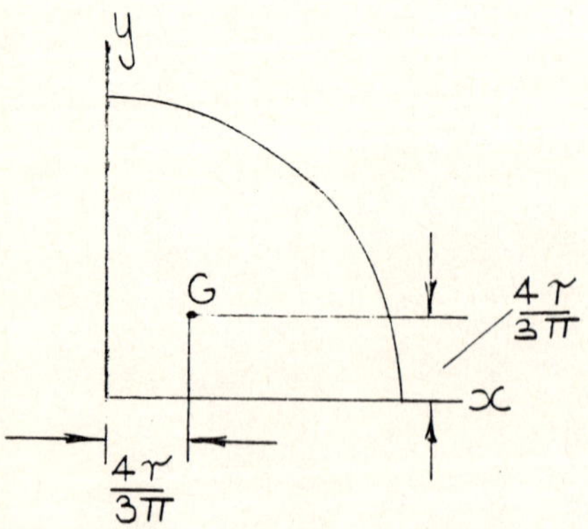

$$I_x = \frac{1}{16}\pi r^4$$

$$I_y = \frac{1}{16}\pi r^4$$

$$A = \frac{\pi r^2}{4}$$

Quarter Circular Area

(b) Volumes and Mass Moments of Inertia

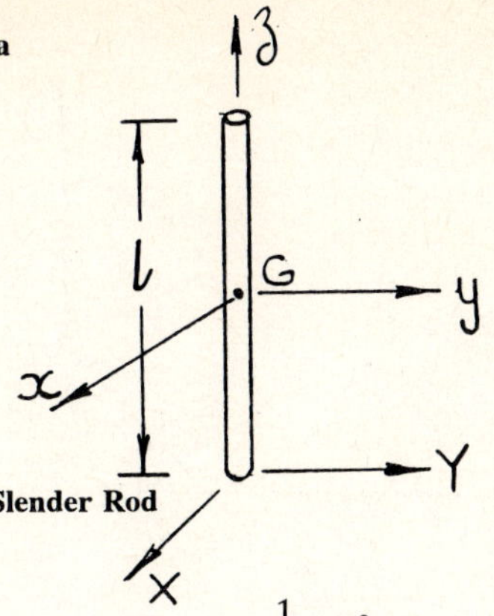

Slender Rod

$$I_{xx} = I_{yy} = \frac{1}{12} ml^2 \qquad I_{XX} = I_{YY} = \frac{1}{3} ml^2 \qquad I_{zz} = 0$$

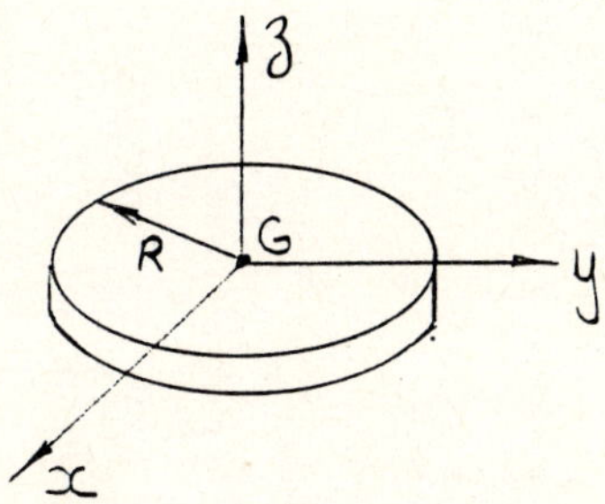

Thin Circular Disk

$$I_{xx} = I_{yy} \frac{1}{4} mR^2 \qquad I_{zz} = \frac{1}{2} mR^2$$

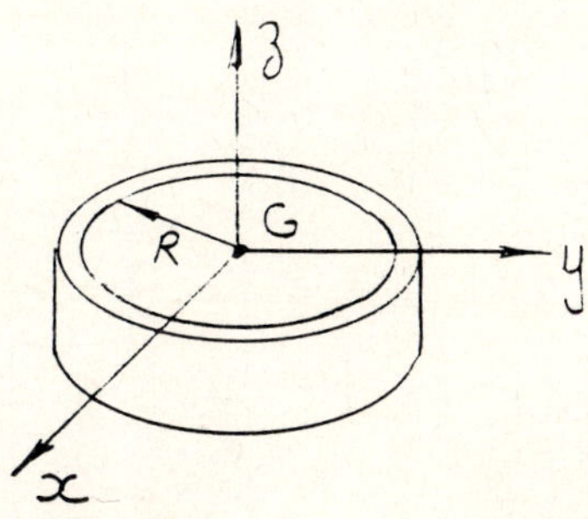

Thin Ring

$$I_{xx} = I_{yy} = \frac{1}{2} mR^2 \qquad I_{zz} = mR^2$$

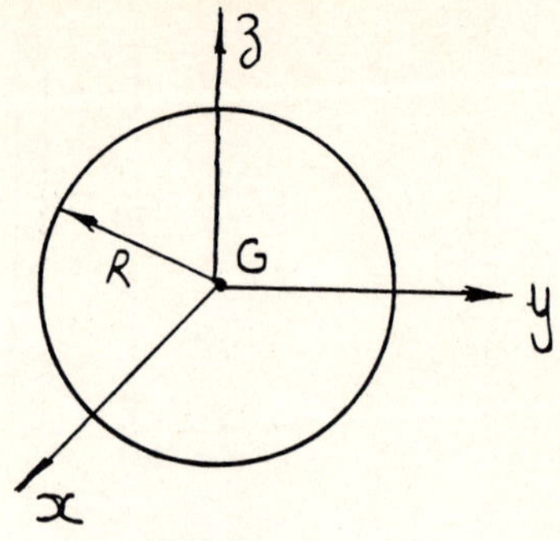

Sphere

$$I_{xx} = I_{yy} = I_{zz} = \frac{2}{5} mR^2 \qquad V = \frac{4}{3} \pi R^3$$

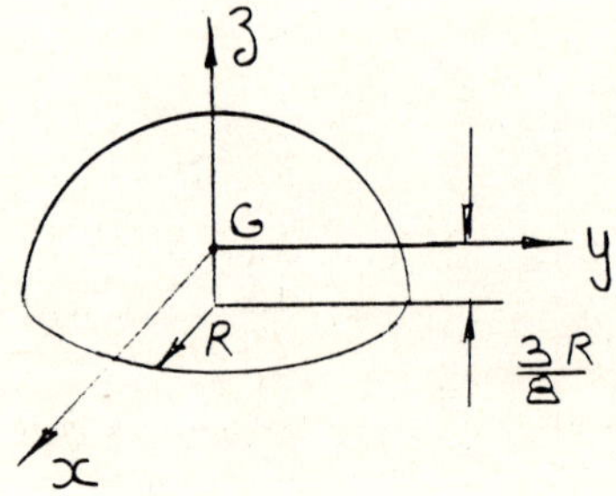

Hemisphere

$$I_{xx} = I_{yy} = 0.259\ mR^2 \qquad I_{zz} = \frac{2}{5} mR^2 \qquad V = \frac{2}{3} \pi R^3$$

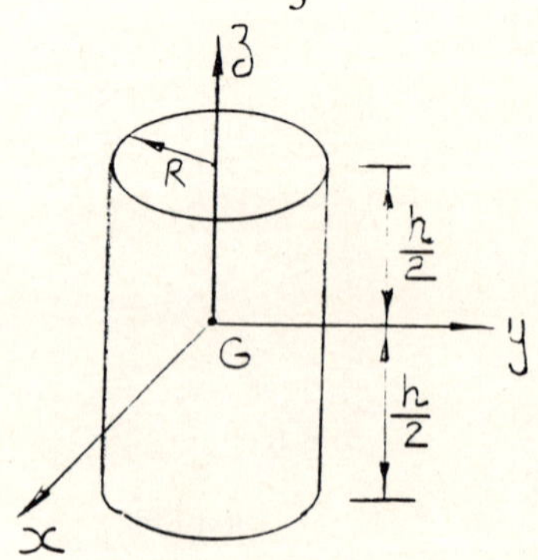

Solid Cylinder

$$I_{xx} = I_{yy} = \frac{1}{12} m \left(3R^2 + h^2\right) \qquad I_{zz} = \frac{1}{2} mR^2 \qquad V = \pi R^2 h$$

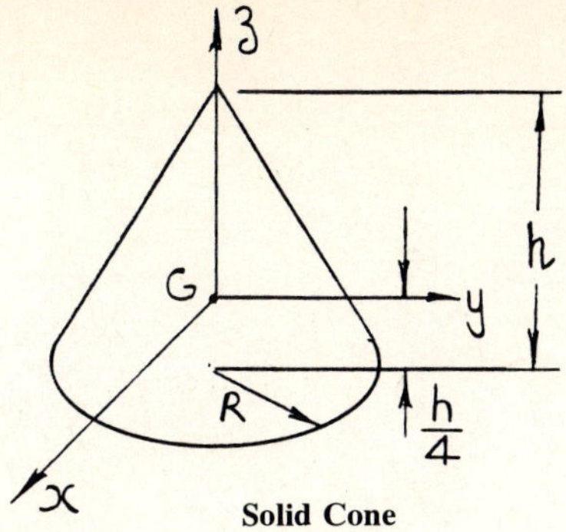

Solid Cone

$$I_{xx} = I_{yy} = \frac{3}{80} m \left(4R^2 + h^2\right) \qquad I_{zz} = \frac{3}{10} mR^2 \qquad V = \frac{1}{3} \pi R^2 h$$

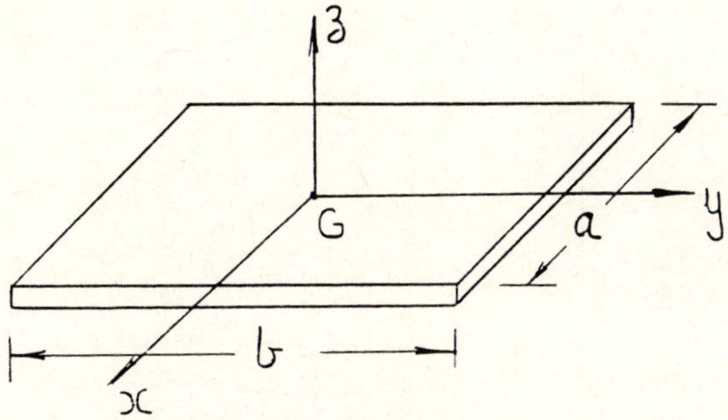

Thin Rectangular Plate

$$I_{xx} = \frac{1}{12} mb^2 \qquad I_{yy} = \frac{1}{12} ma^2 \qquad I_{zz} = \frac{1}{12} m \left(a^2 + b^2\right)$$

INDEX

absolute general plane motion 148
acceleration 1, 19, 20
  angular 31, 140
  centripetal 31
  constant 12
  Coriolis 41, 185
  instantaneous 11
  variable 16

centre of percussion 248 - 250
centrifugal 74
centripetal 31, 74
circular motion 30
coefficient of
  damping 277
  restitution 116
conical pendulum 81
connected particles 55
conservation of
  angular momentum 128
  energy 90
  linear momentum 110
constrained motion 55
Coriolis 41, 185
cosine rule 4
critical damping coefficient 279
critically damped 279
curvilinear motion 135, 137

D'Alembert's principle 74
damped
  forced vibrations 283
  frequency of vibration 280
damping 280, 287
  factor 280
dashpot 277
degree of freedom 265, 290
  multi 290, 292
  two 290
displacements 1
dynamics 5

efficiency 89
elastic strain energy 86

force 1
force, mass, acceleration 62, 199
four bar chain 169
free body diagram 64, 67, 69
frequency 266, 280
friction 67, 68

g 7
general
  planar motion 136, 137
  plane motion 219
gyroscope 254
gyroscopic
  motion 256
  theory 256

Hooke, Robert 87

impact 114
  central 115
  line of 115, 116
  oblique 118
impulse 102, 115
  angular 128, 242
  linear 102, 103, 242
  restitution 116
inertia 192
instantaneous centre of zero velocity 164

kinematics 5
  of a rigid body 135
  of particles 10, 84, 102
kinetics 5
  of a rigid body 192
  of particles 62
kinetic diagram 199, 220
kinetic energy 88, 232

magnification factor 286, 287
mass 1
  moment of inertia 192 - 194
  of Earth 6
mechanical efficiency 89
momentum 102
  angular 126, 127, 199, 200, 242, 254
  linear 102, 103, 242
motion
  about a fixed axis 136
  constrained 55

equations 63, 73
in a circle 30
relative to rotating axes 179

'n' and 't' co-ordinates 28
natural frequency 266, 270, 272
Newton's
law of gravitation 6
law of motion 5, 6, 62
normal and tangential co-ordinates 27, 73, 86
nutation, 257

overdamped 279

parallel axes theorem 195, 196
particle 1
pendulum 81, 269
percussion
centre of 248 - 250
phase angle 268
plane
curved motion 18
curvilinear motion 19
polar co-ordinates 37
position 19
potential energy 89, 235
power 89
precession 257
steady-state 258
projectile 25
Pythagoras' theorem 21

radius of
curvature 32
gyration 193
Rayleigh damping 287
rectangular co-ordinates 20, 63, 86, 103
rectilinear motion 10, 135, 137
relative motion 45
acceleration 171
velocity 153
relative
acceleration 182
velocity 181
resonance 287
rotation about a fixed axes 137, 138, 140, 211, 243

scalar 1
SI 2
simple
harmonic motion 264
pendulum 269
sine rule 4
slider-crank mechanism 168, 178
spin 257
spring 86, 87, 277
mass, dashpot 277, 281
stiffness 87
strain energy 86 - 88
Stroud System of Units 7
structural damping 287

time 1
period 268

underdamped 279
universal gravitational constant 6

vector 1, 3
velocity 1, 19, 20
angular 31
vibrations 264
damped 277
forced 264
free 264
undamped 264
viscous 277

weight 1
work done 84
work, energy and power 84, 231

FINITE ELEMENT PROGRAMS IN STRUCTURAL ENGINEERING AND CONTINUUM MECHANICS

CARL T. F. ROSS, Professor of Structural Dynamics, Department of Mechanical and Manufacturing Engineering, University of Portsmouth

ISBN 1-898563-28-4 650 pages 1996

The finite element method is a most powerful method of solving partial differential equations which apply over complex shapes. Bridging the gap between the expensive and massive software packages and highly academic theoretical texts, this modern-day handbook covers finite element programming in a wide range of problems in mechanical, civil, aeronautical and electrical engineering. Comprehensive, it ranges from the static analysis of two- and three- dimensional structures to stress analysis of thick slabs on elastic foundations, and from two- and three-dimensional vibration analysis problems to two-dimensional field problems, including heat transfer and acoustic vibrations.

The 24 printouts of powerful and valuable engineering computer programs written in QUICK BASIC analyse engineering design and manufactureare problems by the finite element method.

Contents: Forces in Plane Pin-Jointed Trusses: Bending Moments in Beams: Bending Moments in Rigid-Jointed Plane Frames: Forces in Pin-Jointed Space Trusses: Static Analysis of Three-Dimensional Rigid-Jointed Frames: Vibration of Rigid-Jointed Space Frames: Bending Moments in Grillages: Vibration of Grillages: Slab on Elastic Foundation: In-Plane Stresses in Plates: Bending Stresses in Flat Plates: Stresses in Doubly-Curved Shells: In-Plane Vibrations of Plates: Lateral Vibrations of Flat Plates: Vibration of Thin-Walled, Doubly-Curved Shells: Stresses in Solids: Two-Dimensional Field Problems: Solution of Helmholtz's Equation:

Computer Programs: Plane Pin-Jointed Trusses:Beams "BEAMSF": Beams on an Elastic Foundation "BEAMERF": Rigid-Jointed Frames "FRAME2DF": Rigid-Pin-Jointed Plane Frames "FRAMERP" Pin-Jointed Space Trusses: Three-Dimensional Rigid-Jointed Frames: Vibration of Rigid-Jointed Space Frames: Bending Moments in Grillages: Vibration of Grillages: Slab on an Elastic Foundation: In-Plane Stresses in Plates "STRESS4N": In-Plane Stresses in Plates "STRESS8N": Stresses in Double-Curved Shells: Vibration of In-Plane Plates: Lateral Vibrations of Flat Plates: Vibration of Doubly-Curved Shells: Stresses in Solids: Field Problems "FIELD3SF", "FIELD4SF" and "FIELD8SF": Helmholtz's Equation "ACOUSTIC".

"Computer programs for finite element analysis .... .students and lecturers may find them of value"

The Structural Engineer (Professor I.A. Macleod, Strathclyde University, Glasgow)

FINITE ELEMENT TECHNIQUES IN STRUCTURAL MECHANICS

CARL T. F. ROSS, Professor of Structural Dynamics, Department of Mechanical and Manufacturing Engineering, University of Portsmouth

ISBN: 1-898563-25-X 224 pages 1997

This undergraduate and postgraduate text will serve for courses in mechanical, civil, structural and aeronautical engineering; and naval architecture. It is written in a step-by-step methodological approach, so that undergraduate readers can acquire knowledge, either through formal engineering courses or by self-study.

Professor Ross writes in a style easy for comprehension. He is well-known for his many successful engineering books which reflect knowledge and experience in international engineering. He knows the difficulties experienced by most modern undergraduates with mathematics, which he has deliberately made non-rigorous at the level understood in introductory engineering courses. He is well-known for his work on the application of computers in finite element analysis and his important discoveries on buckling of cylinders and cones under external pressure, for which he was awarded a D.Sc. by the Council of National Academic Awards..

Features: Carl Ross's impressive industrial experience strengths the teaching quality of the text, conveys the reality of practical engineering problems, and points to their solution. Worked Examples, Diagrams, Problem Exercises (& Solutions) in good measure.

Contents: Introduction to Matrix Algebra; The Matrix Displacement Method; Finite Element Method; Dynamics of Structures; Elastic Buckling and the Non-linear Behaviour of Structures; Modal Method of Analysis.

"All Carl Ross' previous books are very clear and well written. This thoroughly interesting text is no exception. The worked examples and practice problems are particularly useful. I will continue to recommend Professor Ross' books to my students. For anyone requiring an introduction to finite element analysis, this text is excellent"

Journal of Strain Analysis (Dr. S.J. Hardy, University College, Swansea)

LANDFILL WASTE POLLUTION AND CONTROL
KENNETH WESTLAKE, Waste & Environment Group, Loughborough University of Technology

ISBN: 1-898563-08-X 1995 160 pages

"good information about the science and practice of municipal solid waste landfilling, using a fair amount of data and description ... much of the data will be useful to designers: for example, information on biochemical transformations, leachate composition, gas composition and stoichiometry"

Choice (D.A. Vaccari, Stevens Institute of Technology, USA)

THERMODYNAMICS IN ENGINEERING & PHYSICAL SCIENCE
G.H.A. COLE, Professor of Engineering Design & Manufacture, University of Hull

ISBN 1-898563-22-5 1997 168 pages

"A welcome addition to the library of anyone interested in the provision of energy ... at the heart of the global provision of work ... should be *compulsory* reading ... excellent, lucidly written ... useful information insight in a wide range of disciplines"

(Dr. Jeremy Dunning-Davies, University of Hull)

TEACHING AND LEARNING MATHEMATICAL MODELLING
Editors: S.K. HOUSTON, University of Ulster, Northern Ireland; W. BLUM, University of Kassel, Germany; IAN HUNTLEY, University of Bristol; N.T. NEILL, University of Ulster, Northern Ireland

ISBN: 1-898563-29-2 1997 402 pages

Mathematicians from ten countries contribute interdisciplinary applications in mechanics and engineering, computing science, traffic control, business studies, and mathematics (fractals and analysis). Tertiary and secondary levels covered.

STOCHASTIC DIFFERENTIAL EQUATIONS & APPLICATIONS
XUERONG MAO, Reader in Mathematics, Department of Statistics and Modelling Science, University of Strathclyde, Glasgow

ISBN 1-898563-26-8 1997 384 pages

This advanced course text on stochastic differential equations covers basic principles, with much on theory and applications is also a reference source for pure and applied mathematicians, statisticians, engineers (control and communications), and other scientific areas of application and research.

LINEAR DIFFERENTIAL AND DIFFERENCE EQUATIONS
A systems approach for mathematicians and engineers
R.M. JOHNSON, Senior Lecturer, Department of Mathematics and Statistics, University of Paisley, Paisley, Scotland.

ISBN: 1-898563-12-8 1997 165 pages

An advanced text for senior undergraduates and graduates, and professional workers in applied mathematics, and electrical and mechanical engineering.

"Should find wide application by undergraduate students in engineering and computer science ... the author is to be congratulated on the importance that he attaches to conveying the parallelism of continuous and discrete systems"

Institute of Electrical Engineers (IEE) Proceedings